The Story of Water

Alick Bartholomew

The Story of Water

Source of Life

Floris Books

First published in 2010 by Floris Books

British Library CIP Data available
ISBN 978-086315-738-7
Printed in Great Britain by Page Bros (Norwich) Ltd

For Mari ... *et amicis*

Thanks to Mae-Wan Ho for her inspiration and for generously allowing me to quote from her books and articles; Martin Chaplin for permission to quote from his useful website; Callum Coats for his diagrams; Chris Weedon for his Foreword and helpful suggestions, Kevin Redpath for his encouragement and Christopher Moore for his creative editing; and special thanks to Caroline Way for letting me quote her poem 'Still Water Meditation' which so aptly sets the theme for the book.

Contents

List of Illustrations

Textual Figures:

Colour Plates: (following p.160)

Foreword

I have a great fondness for Water Bears. Less than a millimetre long, Water Bears — or Tardigrades — clamber about like minute animated jelly-teddies in a watery micro-world, on eight stubby legs, tipped with the tiniest of claws. Endearing! But Water Bears are much more special than that. When dry conditions arrive, instead of succumbing to death, Water Bears survive by just drying up . . . completely! Well, almost. Drying to a body content of 1% water, from close to 100%, the creature transforms into a microscopic spec of organic dust, utterly resistant to drought, extreme cold, vacuum and even radioactivity. In this dormant, desiccated state, a Water Bear can survive for thousands of years! It's a good trick if you can do it!

Yet, however remarkable the resilience born of desiccation seems, surely the greater miracle is the life that water brings! For, with even a single drop of water, the sleeping Water Bear bounces back into action, again to stride through mossy jungles. How is it that one extra ingredient has the power to awaken a mote of dust? What has happened? What is water doing? What *is* water that its presence facilitates and empowers life?

The Story of Water sets out in answer to these questions, probing much further than merely repeating that favoured adage that 'Water is Life'. Here is a tale of wholeness and connectivity told through water; of the interplay between material and non-material, enacted on Earth, yet influenced from far beyond the bounds of our planet.

In these pages, the extraordinary subtlety and complexity of water's roles are vividly illustrated. And having done so, the book then asks: what are the qualities of water that best support life? The quest for a comprehensive answer to this question has been the research focus of the 'heroes' of Alick Bartholomew's story. By drawing together the findings and insights of these researchers into so many aspects of water's reality, a picture emerges of a seemingly infinite array of interrelating properties and qualities, which we are only just starting to comprehend. And by analysing and then synthesizing these insights within a single volume, Alick has taken us a step closer to answering that related and most fundamental of questions, 'What is Life?'!

Chris Weedon
Co-Founder of the Water Association
Somerset, UK
February 2010

Still Water Meditation

Place a drop of water in the palm of your hand.

The drop that you hold in your hand
Is part of the water which was the cradle of all life
On this planet Aeons ago
The first rain that splashed down on the hot earth
To form the first sea.
Each drop, in sunlight
Has risen from the sea in countless ages
And fallen to the earth again
As rain

The drop that you hold in your hand
Has been a prism forming myriad rainbows
Has travelled underground streams
Bubbling through dark caverns
The Architect of cathedral caves
Formed valleys
And split granite

The drop that you hold in your hand
Has flowed down broad rivers
Has risen in the sap of trees
Has been the sweat of slaves
And the tears of children
It has become the foam topped waves
And deep unfathomable depths
Of vast dark lakes
And seas

The drop that you hold in your hand
Has been part of the great flood
It has been a dewdrop on a blade of grass
A drop that has been pounded
Through the hearts of whales in blood
And lain in an eagle's egg
It has travelled in the fluid of a poet's brain
And dripped from the wounds of the dying

The drop that you hold in your hand
Has been trapped in the snows of the arctic
Reflected the sun in a desert oasis
And refreshed the weary

This drop
Unimaginably old
Yet fresh and new
Is evaporating slowly from your hand
To mingle with the air you breathe, perhaps
Or drift in a sun-topped cloud
A thousand feet above the earth
Imagine its journey from your hand
Where will it go?
You can direct its journey
As it evaporates
Send your consciousness with it
It is the water of Life
It is still water

Caroline Way

Introduction

Water is good; it benefits all things and does not compete with them.
It dwells in lowly places that all disdain.
This is why it is so near to Tao.
Lao-Tzu

Why is water such an evocative subject? It influences the emotions, the imagination and creativity — artists and poets find inspiration in it. So much in our language is stimulated by water: outpouring, flowing, swell, bubbling, drain, well up. Yet we take it for granted, treating it as a convenience — to quench the thirst, to bathe in and remove rubbish. With the threat that fresh water will become scarce, people are beginning to take it more seriously, but we don't yet understand why living water needs to be treated with care. Water is the most familiar yet, at the same time, the least understood of all substances on Earth.

Water is now in the news because we are told that one of the effects of climate change will be to make fresh water scarce. We are fighting wars over oil, which we worship for its huge energy potential and its ability to create enormous wealth. Wars over water will be even more likely — essentially for survival.

How much more essential to human needs is water, without which we die within a few days! We waste it profligately; not only do we take it for granted, but also the way we abuse it belies its noble character and function. In a real sense, water *is* life, yet we treat it with neglect and contempt. On a personal level, water is the most essential component of our physical and spiritual being, as we shall be discovering in this study. So we *are* water and it is in our genuine interest to understand what is the real nature of this substance. For as you will see, water is a lot more interesting than it seems. Some even see it as an organism with its own life cycle. It has extraordinarily creative roles to play in the creation, maintenance and evolution of life.

Water carries all life. But water is beyond time, for it bears in its flow the seeds of future life, as well as the memory of past life. Water mediates between life and death, between being and not being, between health and sickness. We have lost touch with the magic of water, the freshness of the mountain spring, the reflection in a mountain lake, the mystery of a sacred well. We take for granted that it will come at the turn of a tap. We have allowed water to spread illness and disease. Much is spoken these days of the destructive nature of water. Such water is Nature on the rampage, perhaps showing who is boss, at a time when humanity has wrought so much reckless damage to Earth's ecosystems and to the natural environment.

Sitting on the rocks at the edge of the ocean on a sunny day gives many of us a sense of inspiration and joy. Water is closely linked to the emotions. There is a strong resonance of water with humans. Our ancestors were fascinated by its magic and there is a vast mythology linked to water. Mainstream science does not tell us about this, but we shall be examining research from holistic science that may give us some clues.

You will have heard that our bodies are about 70% water, a similar percentage to that of the ocean surface compared to land. Is this significant? Blood and sap are really variations of water. Earth is known as the planet of water; astronauts wondered at Earth's blue, shimmering watery aura seen from space. But water is believed to exist or to have existed on other planets. Where did it come from?

Water is indestructible. Nature is very good at recycling, and water is what it recycles best of all. As far as we know, the Earth has about as much water in one form or another as it has had for billions of years. It seems as if water was intended for life; certainly life could not have come without it. It has three basic states: solid, as ice, as a liquid, and as a gas; mist could be called a fourth state. There is no other substance that can exist in Nature in these states, within a temperature range of about 100°C.

There are at least sixteen different forms of ice. There are likely to be as many different forms of liquid water; we don't know as yet. There are dozens of different roles played by liquids which are basically water, including blood, sap and about thirty different human bodily fluids, each with its specific purpose.

Caroline Way's poem, 'Still Water Meditation', that opens the book echoes precisely this thought of the indestructibility of water and its many roles in life. But it also illustrates wonderfully that it is through the medium of water that we all share a common heritage — we are all one. Water is the epitome of holism. The mystical naturalist who has inspired my pilgrimage with water, Viktor Schauberger, talked of the whole universe being held in a drop of water.

This study is also a celebration — because it is to water that we owe our very existence. Our aim is to raise the awareness of water as a source of inspiration. The story has not yet been told simply of how water is the stage manager of life, communicating to our bodies' cells how to be part of a vast orchestra, distributing energy in the landscape to make it balanced and productive. Indeed the very laws which govern the harmonious movement of the planets also determine the form and behaviour of our organic life, through water. This extraordinary picture of water's part in the evolution of life derives from the discovery of quantum physics that we are embedded in a vast web of energy that interconnects all of creation. What this book proposes is the novel idea that water and the quantum field are two complementary aspects in the balanced mediation and sustenance of life.

Our biology and physics textbooks tell us that water is merely an inorganic compound through which various chemical processes take place. One of the reasons why we know so little about water may be our obsession with the physical nature of life. Older cultures did not suffer from this limited worldview and consequently appreciated water's special qualities better than we do today.

The poet, philosopher and scientist, Johann Wolfgang von Goethe (1749–1832) was, more than anyone, a bridge between pre- and post-Enlightenment thinking. He sensed the profound difference in outlook between the new rational, mechanistic and more exclusive worldview, and the more traditional inclusive, Nature-centred view. He called the latter 'holistic' science in contrast to 'reductionist' science. Goethe understood that all of life is one, closely interconnected and interrelated, with water as the very symbol of holism, with its role in the sustenance of life.

In the international forum, The Scientific and Medical Network, of which I'm a member, we challenge the adequacy of scientific materialism as an exclusive basis for knowledge and values.[1] We discuss critically, but in an open-minded way, ideas that go beyond the science of splitting and specializing, by integrating intuitive insight with rational analysis and encouraging a respect for Earth and community with a holistic and spiritual approach.

We are here in the boundary zone between the known and the unknown. We shall try to maintain some rigour as to the deductions made from the phenomena we describe. However, the scientific quest has to embrace both intuitive insight as well as proof — otherwise there would be no progress.

Much of what is proposed may seem to some mainstream scientists as mere supposition but, in the interests of advancing understanding, I ask you continually to consider the whole picture — for that is surely what life is about.

The book is divided into three parts. The first part of the book deals mostly with the knowledge of water acknowledged by mainstream science, its anomalous nature, the great water cycles and their purpose — in the oceans, the atmosphere, on the land, under the Earth's surface, in forests, trees, in other organisms and in humans. We also introduce the insights of Viktor Schauberger, the Austrian naturalist, about water as a bridge between the conventional and the new more radical ideas about water. The second part provides insights from holistic science, and running through the whole book is the theme that water is inseparable from a holistic view of life. The third considers aspects of the current water crisis in the world and how we can learn to adapt and reduce its effects.

We shall examine how water is created; its physical and chemical characteristics, and its role in shaping the Earth as we know it today. However, we cannot get far in understanding water by dwelling on its physical characteristics. Water has no identity, separateness or form (the temporary ice crystal being the notable exception). The real significance of water is its role as a medium — for metamorphosis, change, building and recycling, nourishment and information transfer, energy exchange and balancing.

Water has very weird qualities that make it quite different from any other substance or compound. The fact that these anomalies all seem to be weighted towards creating the most favourable environment for life begs the question: Could water be the exception to evolution by natural selection? Could it have been specially 'designed' for our evolutionary potential? This study investigates whether water's extraordinary story might illuminate a quest for meaning.

I believe that the subtle properties of water and its role in the ecosystem are groundbreaking discoveries that place it at the heart of all life processes. Water, it seems to me, could be the handmaiden of that mysterious field of creative quantum energy that surrounds us. Perhaps it could even be called a medium of consciousness! Our forebears regarded water as sacred. Was this just superstition, or did they know something we don't? When Man lived close to Nature, and considered himself to be part of it, there was not yet the feeling of self and other, of separation. The implications of rediscovering this truth through the quantum view of holistic science are profound for the future of human society. This will gradually become clearer as we progress in the latter part of this study.

The book has a radical message: that water as 'the ground of all being', the 'primal substance', is an organism which is self-creating and self-organizing. It governs both life and death. Living, highly structured water is healing and life-enhancing. Debased and polluted water can carry deadly disease and the message of death. There are qualities of water in between which we put up with when we shouldn't. We need to be much more discriminating, and learn how to support life.

Every living thing, from the Earth itself to the tiniest single cell organism, needs both to contain and be surrounded by vibrant water. Water is what brings interconnectedness to all of life. The story of water is a parable for 'all is one', a lesson we urgently need to learn, with the present emphasis on the individual and his wants, and in the prevailing scientific, medical and educational model of splitting things into parts. Water can teach us many things!

Ways of knowing

Science is about knowledge of our world, of life. Its Latin root *scio-* comes from the verb 'to know'. There are different ways of knowing. When Carl Gustav Jung was asked if he believed in God, he replied: 'I don't believe, I *know!*'

The natural world is essentially an indivisible unity, but our present culture is condemned to apprehend it from two different directions —

through our senses (perception) or through our minds (conceptual). A child just observes and marvels, but as our rational minds become trained, we are taught to interpret what we see, usually through other peoples' ideas, in order to 'make sense' of our sensory experience. Both are forms of reality, but unless we are able to bring the two aspects meaningfully together, the world will present nothing but incomprehensible riddles to us.

The Enlightenment was a triumph of reason over received authority and superstition. It brought about an immense advance in knowledge; the modern subjects of physics and chemistry, geology and astronomy were developed.

It also accelerated the severance of our ties to the natural world and put an end to any pretence of being subservient to Nature. This was to result in a philosophy that saw the Earth is basically dead, her resources to be plundered without question. A bias developed towards a more mechanistic worldview, and a distrust of ways of knowing that were irrational.

Our present human society is identified with the conceptual. It is the weakness of the prevailing scientific orthodoxy. Some of the pioneers of science were able to immerse themselves so deeply in the world of pure observation and experience, that out of these perceptions the concepts would speak for themselves.[2] The trained scientist today, however, being burdened with preconceived ideas or principles, is likely to come up with isolated phenomena or at least a very fragmented picture.

Is the universe a machine?

Isaac Newton's *Principia* saw the universe as a machine, and our worldview to this day is based on this premise. Our new technologies are trying to fix the world as though it were an automobile, unaware of the exquisite harmony that holds everything in balance. A machine is predictable because it stands alone and cannot change. A living system is unpredictable, connected to its environment, self-organizing and creative. If I am encouraged to see myself as a separate part of a world-machine, I will feel divorced from my environment, with an inevitable sense of alienation, in denial and with a temptation to escape into addictions.[3] In making our way in the world our heart is pulled in one direction (yes, I can) while the machine-image pulls us in another (no, it's impossible).

In biology, the quick fix is to develop manipulative technologies like genetically modified crops or horizontal gene transfers, regardless of the inherent dangers of pollution and monocultural devastation or the harm to the wider community. One suspects that the priority of GM companies is to maximize profits rather than consider food and environmental quality. Unfortunately politicians seem to get caught easily in this web of self-serving and corruption. In environmental policy, the quick fix is about jumping into unsustainable technologies in order to maintain the profligate lifestyle that we have become used to, rather than trim our way of life to live in harmony with Earth's bounty.

The polarization of science

The enormous energy released in the last two centuries by the exploitation of fossil fuels has fuelled the development of prodigious technological achievements. More crucially, it has given the pioneers of new technologies a sense of dominance over Nature, but also money, power, influence and global spread. Inevitably this worldview has influenced the political and educated classes. It has led to a growth in materialism and the commercialization of values.

On the other hand, discoveries in quantum physics, fractal geometry and the physics of the organism are leading to a new science that is at odds with the orthodox Newtonian theories that are now found to have relevance mainly in the physical domain. This new holistic science can contain the old orthodoxy, but the Newtonian understanding cannot easily cope with the quantum — hence their antagonism. It is as though we are now seeing two incompatible kinds of science.

Before the 1970s, scientific research was funded largely by government. To a great extent this guaranteed its independence. Today nearly all scientific research is funded by business, which naturally has its own agenda that limits the objectivity of the research. This is particularly true in biology, where there is so much profit to be made, but to some extent also in physics and chemistry. The independent Institute of Science in Society is one of few research centres that subscribes to a holistic view of society.

The media's bread and butter is what is new or unusual; but it also thrives on controversy, and the public can get confused, as when climate change sceptics are shown to be credible, or 'bad science' reporters start a witch-hunt against forms of subtle energy medicine such as homeopathy or acupuncture. The cultural climate today is hardly sympathetic to a holistic worldview, yet this is beginning to change as more people realize that the established symbols of authority do not have the answers.

Water and energy

In Part 2, we look more closely at the lesser known, energetic and quantum qualities of water, illustrating how it is able to perform its incredible functions of initiating and sustaining life. We shall study other pioneers of a new understanding of water: Mae-Wan Ho, Theodor Schwenk, Patrick Flanagan, Jacques Benveniste, Masaru Emoto, Andreas Schulz and Cleve Backster.

In discussing holistic research on water's extraordinary qualities, we shall be talking about energy in rather different terms than the mainstream understanding of the term. What is the essential nature of energy? There is much confusion around the term, and if we are honest, we don't really know, except that it always seems to be connected with motion.

Conventionally it means the power to do work and refers to the gross physical energies, such as those produced by a hydro-electric power generator or an internal combustion engine; or a force that produces change. These physical energies are usually termed kinetic energy, or if they are stored and as yet unmanifested, they are called potential energy.

We cannot see energy, only its outward manifestations — its origins lie beyond our senses. There are many extremely high energies which have been measured by science (for instance, nuclear energies), but there also forms of energy of which we are aware but which defy measurement. They are too subtle and cannot be detected even by the most sophisticated instruments. Although science can detect brain activity related to human emotions, it cannot measure their intrinsic power, frequency or vibrational rate (velocity of atomic rotation), nor their true point of origin.

We shall call these dynamic or subtle energies (for instance, quantum energies); they are to do with energizing life processes and appear to operate under different laws than the Newtonian. Viktor Schauberger claims that they respond to the Law of Anti-conservation of Energy,[4] and are therefore conveniently ignored by materialists.[5]

Immaterial or life energies have been recognized and worked with by both indigenous people living close to Nature, and by sophisticated cultures like the Chinese, for thousands of years. One example is *feng shui,* the knowledge of placement in the environment.

The Chinese developed the sophisticated medical treatment of acupuncture with fine needles to correct imbalances of *chi,* the life energy that moves along energetic meridians of the body. Although this practice is widely used in throughout the world by accredited practitioners and by some more open-minded doctors, it is still not recognized by orthodox western medicine. Holistic science, however, is now beginning to identify the nature of the *chi* energy.[6]

Yin/yang balance

The Sun is our main source of energy. In Chinese tradition, it emits a positive, *yang* (masculine) energy. The Earth balances this with a *yin,* or feminine (negative) energy. Polarities are the mechanism of creation and water is their vehicle. The world is governed by the *yin* and the *yang.* They are the essential components for all biological and physical processes and water's working depends on them.

Yin and *yang* are dynamic in the sense that their energy fluctuates — when one grows, the other shrinks. The concentration of energy is a *yin* process, while the tendency to move and disperse it is *yang.*

Western thought tends to think of *yin* and *yang* as fixed states, but in the Chinese tradition they are constantly shifting. Thus in every man there is a woman and in every woman a man, these tendencies varying in different situations. And so it is in Nature. Mornings tend to have *yang* energy, and evenings *yin.*

Fritjof Capra believes the roots of our problems lie in a profound imbalance in all aspects of our culture — in our thoughts and feelings, our values and our social/political structures. In the West we give *yin* and *yang* a moral connotation, seeing them as 'either/or'. The classical

Chinese tradition, however, views them as extremes of a single whole — a constantly changing dynamic balance; only what is in imbalance is harmful.

In Chinese terminology, *yin* corresponds to all that is contractive, responsive and conservative; *yang* to the expansive, aggressive and demanding. They believe that all men and women go through *yin* and *yang* phases. In Western thought all men are supposed to be masculine, creative and active, while women are considered feminine, receptive and passive — a rationale for keeping women in a subordinate role, and for men taking the leading roles and most of society's privileges.

Rather than the Western concept of passive/active polarity, the Chinese is of *yin* as responsive, consolidating, cooperative activity, and *yang* as aggressive, competitive and expanding activity; *yin* conscious of the environment, *yang* of the self. One can see that our society has favoured the *yang* over the *yin* — rational knowledge over intuitive wisdom; science over religion, competition over cooperation, exploitation over conservation. Capra writes:

> Excessive self-assertion, which is characteristic of the *yang* mode of behaviour, manifests itself as power, control and domination of others by force; and those are, indeed, the patterns prevalent in our society.[7]

All processes depend on an unstable reciprocity between extremes. As soon as a process becomes stable it stagnates. It's the same with water; moving, circulating water is energized — still water is effectively dead. Water is the ideal medium for processes because it is an unstable and dynamic medium, and without water, nothing in the Earth environment can change.

With the rising concern over ecology and sustainability there is a profound shift in values taking place — from admiration of large-scale enterprises to 'small is beautiful', from material consumption to voluntary simplicity. They are being promoted by the human potential movement, by feminism, by holistic approaches to health, and by emphasis on the quest for meaning and spiritual dimensions of life.

We shall employ the Chinese use of *yin* and *yang* as tendencies towards extremes of the whole in this part of the book, particularly in connection with Viktor Schauberger's work.

Water retains and communicates energy

Evolution could not progress without the extraordinary ability of water to retain energy. There would be no raising of quality, no healing. The process seems to be tailored to the uplifting of human consciousness, as it is linked to the quality of free will or the ability of choice given to our species.

The health of the body is affected by the quality of your thoughts. Having a positive mind-frame can promote a healthy, balanced cellular health, while anger, negativity and thoughts of self-limitation can result in health imbalances and illness. Our thoughts are extremely powerful.

Our biological water is the medium for all communication, internal and external. Recent biological research shows that, as the 'intelligence' of our whole organism, the water chains allow electrical impulses to reach all parts of the body much more quickly than through the nervous network.

Emotional blocks are a common hindrance to our ability to live up to our individual potential. They can be released through working on the water meridians, through acupuncture, shiatsu or a technique like Emotional Freedom Technique (EFT). See further in Chapter 9.

One of the misconceptions of mainstream biology is that our constitution and our potential are fixed by our DNA. In fact our experiences and how we respond to life significantly affect our personal evolution.

The use of dark field microscopy to track changes in emotional states of the subject may be revealing, as in David Schweitzer's research. Emotions of fear, hate, love and anxiety are instantly shown, reminiscent of Masaru Emoto's research into water samples which record emotional states. (See Chapter 15.)

Perhaps humanity's purpose on the Earth at this time may be to initiate healing and a raising of consciousness in the Earth's soul. I believe that evolutionary advances in times of Earth's upheavals were also made possible by the chaotic situation initiated by the Earth's sense of purpose. But it required water to transmit this new information to create new species and optimization of conditions for a flowering of biological life forms.

This energy-retaining quality of water is the basis of all medicine, whether at the physical, chemical level of allopathy, or at the energy level of acupuncture or homeopathy.

The quantum field

The early pioneers of quantum physics nearly a century ago, working with laboratory microscopes, discovered the strange properties of tiny sub-atomic particles that behave sometimes like vibrant, continually changeable subtle energy rather than matter, defying the Newtonian laws of space and time, and named them quanta.

This promising research was distracted in 1939 by the demands of our society's burgeoning atomic industries and war machines. However, a new generation of physicists has found that quanta do indeed fill the macro environment as well, making an enormous web of interconnected dynamic energy that seems to continue infinitely through space, a kind of communication system.

It resonates with the concept of an etheric substratum from which matter was thought to have been created, which is found in Hindu metaphysics *(prana),* and in the traditions of many early civilizations. This theory attracted considerable scientific credence over the years, but lost its credibility to the materialist worldview in the latter part of the nineteenth century, and you don't now find it discussed in science textbooks.

Is the quantum field the same thing as the etheric field? Not if you go by the current attempts mathematically to interrelate nuclear, electromagnetic and gravitational forces between particles.[8] The quantum field may well be identified with the etheric, but this will require an acceptance that the physical domain is not the only reality. The etheric field in Eastern science is understood to be pure dynamic energy — a difficult concept for Western science. It is just there; it doesn't do anything obvious.

Quantum physics is a new science in process. There is still disagreement about whether the quantum domain may apply to the macro environment — that might open Pandora's Box. The metaphysical implications that can be drawn from the idea of an interconnected field of energy are legion, and one should be wary about jumping to too many philosophical conclusions. Many quantum physicists are sceptical of esoteric and mystical theories. The principles upon which there is general agreement are the uncertainty principle, which put an end to the Laplacian view of a deterministic universe; and that of entanglement, which challenged the idea of completely isolatable systems.

Ancient mystical systems, especially Hindu, Taoist and Buddhist, held water in particular reverence and understood some of its 'quantum' qualities. This is reminiscent of Niels Bohr and Robert Oppenheimer, pioneers of quantum physics, who found remarkable similarities between their new world view and the concepts of the one-ness of all creation held by these mystical systems.

The idea that everything is connected energetically is the foundation of the principle of holism. Water is clearly the common denominator of all life. It is also the exemplar of holism. A discussion of holism can end up as teleology (see below) — anathema to many mainstream scientists who would prefer these heretical ideas to be kept in the laboratory test tube.

I have elected to go with the holistic version of the quantum field, because this is the only theory that seems to support or possibly explain the weird qualities of water. When I speak of quantum water in Part II of this book, I am particularly thinking of those qualities or characteristics that seem to contravene the normal theories of locality and time, in respect to its role in communication and storage of information.

A question of meaning

As far as we can tell, humans are the only order of life that has the ability of self-reflection. We have a need for meaning.

This journey we're taking is a mirror of my own personal search for meaning. When I discovered that water was the key to my experience of life, everything else started to connect, like pieces of a jig-saw. The fascination with water's role in the environment kindled by Viktor Schauberger opened out into a sense of personal connection with the wonders of coherence described by Mae-Wan Ho and other quantum biologists. These interconnections opened up the vast landscape of holism — the view as though from a mountain top. Because of water's multi-dimensional nature, water is the key to a holistic world view.

It is difficult to understand the importance of water through a rational process. When we use our imagination and our intuition the meaning starts to unfold. It is an exciting path, which may illuminate your own vision about the meaning of life. We will start with the nature of the organism, and then open out into the world of subtle energies. I hope you will find the journey stimulating!

> *There is no such thing as a logical method of having new ideas, or a logical reconstruction of this process ... Every discovery contains an irrational element or a creative intuition.*
>
> Karl Popper

PART 1

Our Usual View of Water

1. The Importance of Water

All is born of water and upheld by water too!
Johann Wolfgang von Goethe

The eighteenth century poet and scientist Johann Wolfgang von Goethe (1749–1832) referred to water as 'the ground of all Being'.[1] Thales of Miletus (640–546 BC) also believed water to be infused with Being, believing it was the original substance of the Cosmos. The Austrian 'water wizard' Viktor Schauberger (1885–1958) had a similar view, saying that water is the product of the subtle energies that brought the Earth into being and is itself a living substance:

> The upholder of the cycles which support the whole of life is water. In every drop of water dwells a Deity, whom we all serve; there also dwells Life, the soul of the 'First' substance — Water — whose boundaries and banks are the capillaries that guide it and in which it circulates.[2]

To taste cold, fresh water from a mountain stream is to experience the elixir of life. But what of its sounds, too ...? Do you know anyone who is not affected by the evocative sound of water? Waves crashing on a rocky headland may be awe-inspiring, a tsunami terrifying. But usually the sounds of water are relaxing and healing: the quiet slap of little waves coming up the beach on a still day; the burbling of a brook, or the plip-plop of water dripping in a cave; the swish of the tidal ebb and flow on the beach.

In all symbolic traditions, water is linked with the emotions. It is they that open us out to life; that make us sensitive, receptive and compassionate. Artists love water for its inspiration; it has the ability to stimulate awareness and imagination. Why does water affect us so? Is it not because we are composed mostly of water? it is what unites us; in fact it unites all of life.

The Moon is water's cosmic partner, for the Moon controls the tides. Almost every rhythm, from Moon rhythms reflected in the hydrosphere and planetary rhythms known to biology, right down to the many physiological rhythms found in every living organism, is based on the pulsation of water.

The inspiration of water

As if it were not enough that life is totally dependent on water in all its forms for its creation and sustenance, we are given extraordinary bonuses in the form of the magical beauty that it displays for our wonder and enjoyment. What would life be without rainbows and sunsets, without thunderstorms and cloudscapes, waterfalls, and waves breaking on a rocky shore?

How many great painters and musicians have been inspired by streams, by mighty rivers and the ocean? It is also hardly surprising that water plays a central role in many of the world's religions.

We all need to get away from the daily grind, and when you think about it, many of the ways people relax are to do with water. My cat likes me to take her for a daily walk, and we have a ritual of sitting together on a log by a sharp bend in the river near my home, where the trees on the bank form a tunnel. I watch the clear water's surface while she observes the squirrels and little birds. Now and again a sunbeam pierces the canopy and, reflected off the rippling water surface, makes light waves on the undersides of leaves and on the tree trunks. As the sunlight becomes softer and more golden in the autumn, they are as inspiring as a Mozart symphony.

What a contrast it is to go out on a stormy winter's day to the same bend when the much swollen river is boiling with murky turbulence. Then, at the now impassable ford below my home, a huge loose tree branch is stuck on the depth indicator that shows a depth of two feet. Cars trying to navigate this amount of river have, in the past, been swept downstream.

We forget how inspiring different forms of water can be. For many, winter is a depressing season; but how the land is transformed after a fresh snowfall! If you have been fortunate to have gone skiing, you know how wonderfully refreshing it is to get out on the slopes, whether you like the fast downhill runs, or the more peaceful gliding through trees on an unbroken surface of powder snow. One poignant memory is of the winter stillness of a New England pond, the swishing of my ice skates reflected by the trees lining the banks.

When I lived in Boston, I used to go down to the Charles River basin after work and sail a little Mercury sloop. Now that was relaxing, even though I was still in the big city! Sailing is a great way to let go of life's worries. You are at one with the wind and the water. My niece sailed round the world in a small boat to settle in New Zealand. What a character-building experience that was!

I first crossed the Atlantic by sea in a troop carrier in 1948. It was fun later to go on the big Cunard ships, before commercial flights got going; the ocean helped one to lose the sense of time. But what I really preferred was to go on a small cargo boat, when one could spend all day up in the bow, watching the dolphins or the long-distance seabirds. The ocean is mesmerizing, calms the busy mind and stimulates your inner philosopher.

Surfing provides the most intimate contact possible with water. It gives the rare opportunity to be at one with the water and its energies. The vast curling tunnel of the mega wave creates a powerful energy vortex down its centre.[3] Aficionados say that 'going down the tube', when the surfer is totally enclosed by a fast moving tubular wall of liquid energy, is the ultimate experience (Plate 8). Or 'hanging ten' when the board is cresting a big breaking wave, and the surfer feels as though he's going over a precipice. The surfer's mind is lifted by the vortex onto a higher energy level for a timeless moment that can give a sense of one-ness with the whole ocean and with all of Nature. To experience this even once can be a life-transforming experience. This may be the reason why many experienced surfers become mystics.[4]

The common thread in all these experiences is that they give a sense of timelessness, but also of connection with a natural energy that is more specific than that of the natural environment in general. It also gives us a sense of 'the spaces in between' and makes us yearn for solitude and silence. Water is an intimate part of the human on a physical level, but yet perhaps even more so on psychic and spiritual levels. As we shall examine later, water gives us a connection with the Cosmos.

What is water?

As we are composed mostly of water, it ought to be the most familiar substance imaginable. We actually know very little about it. It is the most mysterious substance on Earth; even the new sciences throw little light on it. One difficulty about grasping the whole nature of water may come from our intimate physical relationship with it. As we *are* water, it is hard to form a detached view about it!

Second only to hydrogen, water is the most common molecule in the Universe and fundamental in the formation of stars. It is found as widely dispersed gaseous molecules and as amorphous ice in tiny grains, as well as in much larger asteroids, comets and planets, but water needs particularly precise conditions in order to exist as a liquid as it does on Earth. It is thought that this water was transported to Earth by comets and asteroids, and also arose out of the interior some time after Earth acquired a crust.

States of water

There is no compound other than water that can exist in its three basic states within such a narrow band of temperature and pressure. Each state has its vital and special purpose: ice, for weathering of rock, is the most stable state; liquid water, for energy transmission in the earth and in organisms, is chameleon-like and sensitive; vapour, for driving the atmospheric and greenhouse systems, is the most unstable.

We tend to think of each state of water as specific and lasting, but there is a constant interchange between them. When we consider that one of the principal features of life is constant change and transmutation, this is because of these qualities of water.

Water's quality of absorption is absolutely crucial to all processes on Earth. In a sense, the separation of ocean from atmosphere is arbitrary. Water is constantly evaporating from any body of water, leaving impurities behind, forming the important atmospheric gas, water vapour. Mist is tiny droplets which are seeking bits of dust in order to make larger drops of water. When you go out walking in a thick fog, your coat allows the mist to condense as wet water.

Water also absorbs gases, a very important aspect of the greenhouse effect. Its ability to contain gas depends on temperature — the cooler, the more it can hold. (This is opposite to the atmosphere, which can hold more water vapour if it is warmed.) The principal greenhouse gas is carbon dioxide (CO_2), which is absorbed by oceans and forests, as 'carbon sinks'. We do not yet understand the mechanism by which the oceans normally increase their rate of absorption of CO_2 when the level of this gas starts to raise the greenhouse effect above a balanced level. Unfortunately we are now in abnormal times and global warming is reducing the efficiency of forests and oceans as carbon sinks.

Methane is a crucial greenhouse gas because it is twenty times more powerful a shield than CO_2. When the Arctic tundra melts, masses of methane and CO_2 are released. There are enormous amounts of methane hydrides locked up in the ocean shelves, which could be released into the atmosphere if the temperature of the sea rises to a critical level. All these changes influence each other as positive feedback effects, creating magnified outcomes which we are only just beginning to appreciate.

Ice also absorbs gases, and information on the oxygen content of the atmosphere in prehistoric times can be gained from analysing oxygen bubbles in ancient ice fields. Recent research shows that oxygen levels are being depleted faster than those of CO_2 are increasing.[5]

Water is a neutral medium

Water is an unselective host body for nutrients and pollutants alike. Holistic biology reports that water is also a vehicle for storing and communicating subtle energies, both enhancing and destructive of life. The quality of the water determines which role it is able to play, and this is governed by general environmental factors, by the shapes and forms of organisms and by the particular response of water to temperature and certain types of movement.

Pure water exists in Nature only in evaporated form. This is not healthy water; in order to become so it must acquire its complement of minerals, salts and trace elements. Radical scientific thought believes it is then 'mature' and ready to perform its role as the nourisher and sustainer of all life.[6] But water is easily degraded, and we need to understand how to help water retain its life force if the natural environment and we are to be healthy.

Because it is so familiar, we regard water as a 'normal' liquid. It is anything but normal. The ways in which it is quite different from other compounds are called its 'anomalies' (see Chapter 3 below) which suggest that water was intended specifically to bring forth life.

Water is the most fundamental active ingredient in Nature's processes. Without it there could be no life. Life cannot evolve or be sustained without liquid water, which is why we want to find water on other planets and moons. Enzymes — the proteins which increase the efficiency of biological cells — cannot function without water molecules.[7]

Even though we drink it, wash and cook with it, we inevitably overlook the special role it has within our lives. We need to have enough — too little causes famines, or excess of the stuff causes disease and death; both droughts and floods are killers. When the environment is in balance, water is a healer and the nourisher of life. When out of balance, water can become the bearer of death, bringing deadly disease.

The most important of water's anomalies are those that orchestrate the temperature range of liquid water (0° to +100°C). This range is wider than the liquid states of other similar compounds. It neatly encompasses the range that is required by organic life. Clearly life on Earth depends on the unusual structure and anomalous nature of liquid water.

The properties of water are so unique and clearly adapted to the requirements for life that the living world should be seen as a partnership between biological molecules and water. It performs so many functions that you might even think it *is* life. It energizes, nourishes, transports, lubricates, reacts, stabilizes, signals, partitions, structures and communicates.

Water as facilitator and stage manager

Without doubt water is the most important substance on Earth. It drives everything, from the most delicate metabolic processes in our bodies, to creating environments favourable to life, to weather patterns and climate change.

Dynamic water, when it is alive and energized, performs the roles of initiating and operating all the processes of life. The most important function of biological water is to facilitate rapid inter-communication between cells and connective tissues, so that the organism can function as a coordinated whole.

Though not recognized by mainstream science, living water in fact performs this inter-communication function between organisms, groups of organisms, populations, natural kingdoms and worlds, creating a network of sensitivity throughout all of life, so that nothing can happen without affecting other processes; all are linked together by water. In this way, it drives evolution.

But its workings are ambiguous. On the one hand it seems to initiate processes and yet its role is the more passive one of facilitation. This should become clearer as we progress through this study. It is similar to a good doctor or healer denying that they initiate the healing process — rather playing a role of facilitating the subjects' own potential for healing.

Our incomplete knowledge of water's role and function makes it hard to predict processes of instability and change that we are experiencing today. Water magnifies and accelerates a process like that of global warming (or cooling). It is vital that we understand water's role in the critical tipping points of the warming cycle, especially positive feedback loops. Water's role is neutral; its influence can be either beneficial or destructive. In global warming its role is fundamental and needs urgently to be understood.

Water's shape

The ideal form for water is the sphere. This shape gives water integrity and allows it to circulate and retain its energy (though it does this best in an egg-shaped drop). Water is alive only when it moves and is able to develop layers and filaments. These structures are invisible unless you add something like potassium permanganate to show them up.

When the drop hits the ground its shape is lost, but it will seek a slope down which it can develop these structures in its flow. A good place to observe water movement is on a smooth, inclined road surface. Then you can see how it is constantly changing direction, back and forth, like a dance. On a beach a stream will erode the bank on the outside bend and deposit sand on the inside.

The strangest thing you notice about water is that it moves rhythmically. On the slightest slope the water's surface starts to move and then it becomes filled with surface-like plates with vortical structures. Indeed water is the element of movement; it is a carrier. This movement is its function and its magic.

As soon as it is moving it can perform its real function, which is to open up to the environment; in movement it fulfils its potential, which is to bring life. Water sacrifices itself entirely to its surroundings. Everything in Nature depends on water. Movement is influenced constantly by its very presence.

Caring for the stream

What is really fun to do is to wade into a small stream on a summer's day with a butterfly net, disturb the gravel and mud on the stream bottom with your feet, let the net sieve some of this water, and drop the contents of the net into a tray of water. You will be astounded by the variety of invertebrate life, from tiny shrimps to insect larvae to worms of many kinds. The presence of dragonfly and damselfly larvae indicates a healthy stream. They spend 6–7 years as larvae in the mud of the stream's bottom, and on pupating have only several weeks in the sun.

With control of pollutants, the water quality of our rivers has improved enormously in recent years, and the conservation authorities are trying to encourage a wide biodiversity of fauna in rivers. Typically there may be thirty family groups of invertebrates, some of which may contain 200–300 species. This is a sign of good water health, which enhances the number and variety of fish and brings back animals like otters and water voles.

Viktor Schauberger, the Austrian naturalist, claimed that streams attract trees to their banks to keep the water cool.[8] Through careful study of the ecological balance, sensitive management schemes can be put in place to protect their eco-systems and maximize their biodiversity.

The sea shore

The inter-tidal seashore is hybrid territory and its fauna special. There are the rocky pools with a fascinating life of little crabs, mussels and worms. The tangle of bladderwort and other kelp — with their attendant insect life, abundant when the sun comes out — creates a pungent aroma. It makes a rich compost, traditionally used in the Scottish Hebrides and the Channel Islands as a fertilizer. During the Second World War there was a seaweed research laboratory in my village, creating products like a special powder surgeons used inside their rubber gloves.

The sandy beach looks quiet and peaceful, but under the surface are insect larvae, and worms which throw up their casts in little spiral pyramids. And when the tide retreats, flocks of wading birds are attracted to the rich takings.

I never did like building sandcastles that much, preferring to divert the little streams coming down on the beach into pools and harbours. My favourite pastime, however, was searching for the elusive little pink cowry shells, which still fill jam jars at home.

Perhaps this special environment reminds us that part of us still belongs to the sea. Certainly it draws us, and many would regard the seaside as the only proper place for a relaxing holiday.

Listening to the stream

Indigenous peoples, who lived closer to Nature, understood the importance of water and how to look after it, venerating it.

One of the principal cultural ceremonies of the Pacific North-West indigenous tribes in British Columbia was the 'potlatch', believed to have been practised for thousands of years. Usually a part of this complex ceremony was given to the recitation of the spiritual traditions of the tribe. The 'Speaker', a position often handed down in one of the chief families, performed this important role.

The training of future speakers was a long-observed ritual, involving teaching the young person to listen to the wind from mountain tops, to the sound of the ocean waves on the rocky shore, and to the music of the rushing stream. They would in this way learn to incorporate Nature's sounds and cadencies in their telling of the stories of the tribe's wisdom. To these people, water was a sacred medium of communication and they felt its sounds should be the vehicle for teaching their traditions.

How water informs language is hinted by many of the ancient scripts, which employ flowing lines like water in their characters (for instance, Hebrew).[9] Greek became more regular, but it was Latin with its angular, straight characters, that forgot the memory of water.

There are esoteric uses for water. Nostradamus used a bowl of water as a skrying tool for seeing future events. Viktor Schauberger had a remarkable experience while sitting by a rushing stream in his pristine Alpine refuge. Listening to its vivacious music, he intuited how water needs to move and behave in order to stay healthy, which was to inform the groundbreaking research that earned him the title of the 'Water Wizard'.

How we regard water

Until quite modern times, water was always regarded as sacred. The Greek and Roman civilizations still appreciated that water had special qualities. Up to the eighteenth century there were those, like Leonardo da Vinci and Goethe, who understood that water is an organism that is alive, and that it promotes life.

Our attitude to water has changed enormously in the recent centuries. This precious substance used to require a great deal of effort to collect for domestic use (and unfortunately still does in some countries). It was treated with reverence and was believed to be protected by divine beings. With the advent of rationalism and the denial of spiritual influences on humanity came the great explosion of technology that loudly proclaimed humanity's supremacy over Nature.

Since we decided we were not part of Nature and made up our own self-centred laws, we have lost touch with the magic of water. We have forgotten its true nature and the meaning of its pulsating movement.

Water in a state fit enough for human consumption or for succouring the life cycle of the brown trout is now in short supply and is diminishing every day. Though we may possibly be able to live a more balanced life without oil, without good water we shall soon perish.

The key to understanding water — indeed to start living more in tune with our environment and with Nature — is to learn to see and feel holistically: that we as individuals are part of a community of other beings, human, animal, microbial and botanical, united by the common bond of water.

But we are forgetting that water is universal. What are its origins and its history?

Water is the source of all life.
Attributed to Thales of Miletus (634–546 BC)

2. The Cosmos and the Solar System

God is a mathematician, and the universe is beginning to look like a big thought rather than a big machine.
Sir James Jeans, astronomer

Water in the Universe

The main product of the Big Bang was hydrogen (H_2 –twinned atoms), which is still the most plentiful element in the Universe. The prevailing theory of water's origin is that the process of star creation broke down and re-formed the primitive elements into oxygen and all the other elements that make up our world, including water (the word 'hydrogen' means literally 'water creator'). As the gas collapses to form a new star, it appears that water is born.

Forty years ago, scientists had only one solar system to study — our own. A recent revolution in astro-technology means astronomers can now spot Earth-like planets orbiting faraway stars, raising the chances of extraterrestrial life being found. In the past twelve years, more than two hundred planets have been detected outside our solar system by measurements of tiny gravitational wobbles of distant stars. Many of these show the presence of water.

In some parts of the galaxy, the free gas and dust between the stars form opaque clouds that block out the light (like Orion's Horsehead nebula). It was in one of these, in 1969, that Charles Townes discovered a bright peak in a microwave spectrum of the cold interstellar gas that indicated an abundance of water. It seems that there is plenty of water in these star nurseries as ice, or as steam, but seldom as liquid water.

James Lovelock, proponent of the Gaia hypothesis, coined the term 'the Goldilocks Zone' (also called 'habitable zone') for the necessary planetary conditions such as size or temperature to be 'just right' in order to sustain life (from the story of Goldilocks, who preferred porridge which was 'not too hot, and not too cold'). *Liquid* water, the requirement for life, can exist only in the Goldilocks Zone, according to this theory.[1] Most of the recent planetary discoveries show that, so far, the telescopes have been able to pickup only the gas giants that don't meet the Goldilocks criteria in terms of size and temperature.[2]

What is extraordinary is that the temperature range between 0° and 100°C that is suitable for life is the same as for liquid water and for Earth's temperature range. This range is less than 2% of the average found among the planets in our solar system. Although our Sun has cooled considerably since Earth was formed, the Earth has warmed the same amount in compensation, due to the heat distributed by water.

Water in the solar system

Our solar system is thought to have formed from a nebula over five billion years ago, some of the pristine remnants of which are found in its outer reaches. The Oort Cloud — believed to extend from 50AU to 50,000AU (astronomical units)[3] — consists of debris left over from the condensation of the solar nebula and probably trillions of comet nuclei. These objects are composed mainly of ice (predominantly water, but some methane and ammonia), and represent a gigantic reservoir of water.

So how did Earth's oceans become filled? The solar system, when it started to coalesce, would have contained a great amount of water which would be distributed amongst the bodies that condensed out of it. Proto-Earth would have released this interior water as a vapour held near the surface by Earth's gravitational field. The mythology of Earth creation speaks of rains lasting centuries.

For a billion years after Earth's formation, comets were extremely numerous. There would have been a massive bombardment of Earth by these wanderers. Typically a comet is several hundred feet to several miles in width and composed principally of ice. It would take half a million comets hitting Earth to provide half of the oceans' contents. Cometary water has a high content of heavy, deuterium-rich hydrogen, whereas oceanic water has far less deuterium. When Earth was formed, the early magmas contained a lot of water. Rachel Carson suggested in *The Sea Around Us* that about half the oceans' total would have come from deep inside the Earth. More recent theories suggest that new primary water is constantly being produced in the Earth's mantle.

> Primary water is generated in the rock strata when the right temperature and pressure are present. It is then forced into fractures and fissures in the rock where it can traverse large distances, hundreds of kilometres. Some of this water is expressed as springs which can be either hot or cool. This water is always moving and therefore can be detected by dowsing.[4]

A major source of comets is the Kuiper Belt, which extends outwards from Neptune, to about the orbit of Pluto. There are thought to be ice-encrusted objects also in the Asteroid Belt, between Mars and Jupiter. There is no shortage of frozen water in our solar system, and it seems to have been present since the beginning of time — but liquid water?

Comparison of photographs of the surface of Mars taken in 1999 and 2006 by orbiting satellites showed recent deposits in two gullies that could have been made only by flowing water. It is now thought that the planet may have large underground reservoirs of water that occasionally gush up as aquifers.

Astronomers have found evidence of liquid water on Saturn's moon Enceladus, a discovery that raises the possibility that it could support life.[5] Images from the Cassini spacecraft showed that the South Pole region was geologically active with erupting plumes of water.

Why is Earth the only planet to have oceans? There is a narrow temperature range that allows liquid water to accumulate. Mars is the only other planet in our solar system within that range. It is probable that it once carried oceans on its surface. Its smaller mass and lower gravity would have made it difficult for Mars to hold on to its atmospheric water. However, it is now thought that Jupiter's planet Europa may have abundant water *below* its icy surface.

A study of the links between physical processes in the Earth and life can help an understanding of evolution. The Earth is generally thought to be about 4 $^{1}/_{2}$ billion years old. The first two billion years are clouded in mystery. But let us say that it would have taken about $^{1}/_{2}$ billion years for the planet to develop a crust and to start to form oceans. Probably about two billion years ago, the oceans became the womb for primitive forms of organic life, initially bacteria, worms and algae. It was impossible then for life to gain a foothold on land which, up to 400 million years ago, had a most inhospitable environment.

There was no climate as such, no fertility. How and when the atmosphere formed is also a mystery. The early atmosphere must have been composed mostly of CO_2, creating a very hot surface temperature (like Venus). It is thought that a cosmic impact (for instance, when the Earth/Moon system was created) may have removed this heavy atmosphere, to allow a thinner, life-friendly one to develop. Once photosynthetic plants became established on land and micro-organisms in the oceans, the formation of an oxygen-rich atmosphere would be able to proceed.

The Earth-Moon system

We cannot understand water without accepting its close relationship with the Moon. They are so intimately intertwined. All ancient cultures celebrate this. In many traditions the Moon has also had a strong influence on the understanding of life cycles and on planting times for crops.

There are many theories of how the Moon was created, but most agree that its age is similar to Earth's. There are two principal theories for the Moon's origin. One is that it was formed as the result of a collision of a Mars-sized object with a very young Earth; the other that it was formed by an accumulation of left-over debris, just after Earth's formation.

The evolution of terrestrial life on Earth as it has unfolded would not have been possible without its extraordinary partnership with the Moon. There is no other planet-moon combination known where a moon is as large as ours is in relation to its planet (one quarter the size of Earth — it has been called a 'binary planet' system). The generation of Earth's strong magnetic field, which lessens damage to life systems from cosmic rays, was probably due to prolonged heating following the likely impact that created the Earth/Moon system.

The Moon has strongly influenced the evolution of terrestrial life

— primarily by stabilizing Earth's obliquity (inclination of its axis to its plane of rotation) at 23°27', which allows predictability of seasons, climate, weather patterns, plant and animal environments *(cf* Mars's obliquity varies 0°–60°). It is possible that the stability of the Earth/Moon system has also limited disruption from cosmic impacts, always a threat to the development and maintenance of life. Clearly, Jupiter and Saturn, because of their enormous masses, have been the first line of defence for Earth life (for instance, at the time of the Shoemaker-Levy cometary impact of Jupiter in 1994), but our Moon is an inner defence system.

The Moon's ability to generate large tides created a littoral environment with tidal pools to encourage polymerization of organic molecules and would have facilitated the emergence of life from the ocean onto the land. The effectiveness of water as the driver of life and evolution for all organisms is facilitated by the tidal influence of the Moon. The deeply rich diversity and sheer abundance of life as we see it on Earth is largely a product of this special Earth/Moon system. The Moon's gravitational effect on water in organisms is clearly extraordinarily important in promoting growth patterns. (See Chapter 13.)

The inevitable conclusion is that it is hard to see how other planets fulfilling the Goldilocks criteria could produce organic life *as we know it* without a moon like ours.

Gateways in time

Geological periods are quite distinct (see time chart in Figure 1). They are typified by specific geophysical conditions, climates and life-forms. Very often the opening of a new period is heralded by one of four agents of change: the shifting of continents (with mountain building); cosmic collisions (for instance, the demise of the dinosaurs); climate change driven by changes in CO_2 (for instance, the end of the Permian period) or worldwide glaciation. These gateways often caused species' extinctions, but they bring in new species or life-forms with an escalation in complexity, which is an essential part of Earth's evolution towards higher quality or 'consciousness'.[6] They can be seen either as setbacks or as times of great opportunity for life. It seems that Earth's evolutionary path is closely connected to the evolution of life.

Eras	Periods	Million Years	Earth Movements and Glaciation	Life Forms
QUATERNARY	Holocene Pleistocene	2	(glaciation)	6th mass extinction man
TERTIARY	Pliocene Miocene Oligocene Eocene	65	(ice caps) ALPINE (vulcanism)	higher mammals modern forests bony fish
	Cretaceous	144		5th mass extinction (dinosaurs) reptiles flowering plants ammonites
MESOZOIC	Jurassic (New Red sandstone)	205		more diverse forests first mammals and birds dinosaurs
	Triassic	248		4th mass extinction coniferous forests
	Permian	290	HERCYNIAN (glaciation)	3rd mass extinction (oceanic)
UPPER	Carboniferous	354		coal forests first land vertebrates sharks
PALAEOZOIC	Devonian (Old Red Sandstone)	417	Late CALEDONIAN	2nd mass extinction primitive trees first plants and animals soils form climatic zones
	Silurian	443	Early	
LOWER	Ordovician	495		1st mass extinction first fishes
	Cambrian	545		Cambrian 'explosion' abundant marine life esp. trilobites
	Vendian	650	CADOMIAN	precursor to explosion of oceanic life forms
PROTEROZOIC	Cyrogenian	1bn	(Snowball Earth) glaciation	ocean salinity present level primitive oceanic life *Eukaroytes* bacteria
PRE-CAMBRIAN		2.5 bn		
ARCHEAN		3.5 bn 4.6 bn		oxygen reltd blue-green algae *Extremophile prokaryotes?* Moon lavas (maria) oceans forming oldest rocks crust forming Moon's cosmic bombardment Earth's and Moon's births

Figure 1. Geological time chart. The boundaries between geological periods signal a marked change in the Earth's environmental conditions. More often than not, they are connected with climatic changes like glaciation or global warming; sometimes by an asteroid strike or by orogeny or vulcanism caused by plate tectonics. We call these periods of Earth restlessness, and they have caused the five great species' extinctions which were followed by evolutionary explosion. (A. Bartholomew)

A dozen or so causes of mass extinctions have been proposed, including super volcano eruptions, change in sea level, meteor collisions, and sudden or lasting cooling or warming. There is as yet little appreciation of the mass extinction that is now under way. There are millions of species on the Earth, but 70% of the terrestrial ones are concentrated in only thirty-four biodiversity hotspots, which used to amount to 15.7% of the planet's land area. These small and declining patches shelter many species that occur nowhere else: at least 150,000 endemic plant species, almost 12,000 endemic invertebrates and many millions of invertebrates, mostly unknown to science. The terrible judgment on our species is that 86% of this habitat has been destroyed, mostly since 1950.

It is not possible to slash ands burn 86% of the habitats of tens of millions of species without at least half of them becoming extinct. Not necessarily at once, but committed to extinction they will be, due to the reduction and fragmentation of their habitats and the death of partner species such as their pollinators and seed-dispersers. If the whole dynamic were stopped today, we would still be looking at millions of species continuing to die out, probably at an accelerating pace as the struggle ends for thousands of ecosystems. This process seems set to peak in the period 2000–2025, when half the world's species are likely to be lost, at a rate of about a million a year.[7]

Mountains, eroded by ice and water, provide the nutrients for evolutionary advance. There is little evidence of cosmic collisions far back in the record but undoubtedly there were a number, some of which might have been responsible for gateways in time. We know only of more recent ones where the evidence still exists on the Earth's surface, such as the extinction of the dinosaurs about 65 million years ago at the end of the Cretaceous period. There is also evidence of one much more recent, about ten thousand years ago near the end of the last ice age, which may be connected with the myths of the Great Flood.[8]

Glaciation

The most recent theory about the global thermostat is that it is controlled through the action of CO_2. The global thermostat takes care of most variations of the heat received from the Sun. The planet has been ice-free for 90% of its existence. However, for at least the last thousand million years, the Earth has switched between a balmy greenhouse climate and one when significant amounts of ice covered Earth's surface.

There are two factors which may have contributed to the onset of glaciations. First, the Earth has an elliptical orbit around the Sun; when it is furthest from the Sun, it might cool sufficiently to have ice sheets form. Second, the continents move on the surface of the globe, and if they were to group near one of the poles when Earth is furthest from the Sun, this could cause a worldwide glaciation.

This has happened at least four times and, as we have noted, they seem to have stimulated evolutionary forward leaps. Ice caps at the poles result in a world climate balance that seems to favour biodiversity.

Earth restlessness

At the dawn of the Palaeozoic era, some 545 million years ago, Earth went through an evolutionary advance, which produced an explosion in the biodiversity of ocean life-forms, when quite complex organisms started to appear — vertebrates and hard shelled creatures like the trilobites. An abundance of oceanic fossil remains is found in the geological strata from this time onwards.

Then, about two hundred million years later, half way through the Palaeozoic era (400 million years ago), Earth experienced another great restlessness — the Caledonian earth movements. These raised great mountain chains extending from Ireland, through Scotland to northern Scandinavia, as well as in other parts of the world, whose land masses were then differently positioned.

One might describe Earth's four billion years before this as a long period of gestation, with the gradual cooling of its surface, establishment of solid crust with continental roots, formation of oceans, balancing the chemistry of water and the creation of a primitive atmosphere and, eventually, experimenting with basic life-forms in the womb of the oceans.

Life emerges from the ocean

The first terrestrial life-forms were types of seaweed that absorbed CO_2 and produced oxygen, and mangrove-type semi-aquatic plants and primitive ferns. Then the first non-aquatic plant life appeared, enabled by the creation of soils that could later support the first forests in the Carboniferous period and which allowed the first land vertebrates to emerge — amphibians and early vertebrates. The chemical composition of seawater creates an environment that limits the evolutionary potential of sea creatures. The ocean mammals got around this much later by developing a heart and blood circulation that required oxygen.

Life has had a more tenuous hold on the dry land of this planet than would appear from the apparent success it has enjoyed in the current age. However, during the mass species' extinctions of the Ordovician, Devonian, Permian and Cretaceous periods (see time chart in Figure 1), oceanic species were badly hit by the over-acidity of the surface oceans.[9]

Earth as an organism

When James Lovelock first proposed his Gaia hypothesis about 1971, mainstream science castigated him as a romantic. However, the concept of Earth as an organism is really quite ancient. It had been given scientific credence by Viktor Schauberger in the 1930s, but his ideas remained in the German esoteric field until the 1990s.

The story of Earth as a living, breathing, organism that is constantly evolving and maturing, being able to sustain life-forms of ever-increasing complexity, mainstream science still finds very hard to accept. This understanding shows that Earth is able to regulate the biosphere's environment, controlling temperature, renewing its skin. It is a remarkable story that, even when faced with cataclysmic events, Gaia demonstrates the ability to heal itself.

The Darwinian view is that life evolves by selection in the face of environmental influences, whereas the Gaia hypothesis claims that life influences its environment in such a way as to optimize its own future.

It seems as though Earth may actually create these events in order to stimulate evolution. We now accept that forest fires help biodiversity; in the longer scheme of things, orogeny (mountain building), continental lift and submersion help to rejuvenate fertility.

There is much talk today of the need to 'save the planet'. Earth has brilliant resilience to adapt to change. Today's threat will be no exception. It is humanity that is facing its own self-destruction.

Ninety per cent of life is in the oceans

Life originated in the oceans, some 3,600 million years ago. Beneath the Earth's surface down to a depth of 11 kilometres, millions of species and countless ecosystems flourish. It is easy to think of life on the Earth primarily as a terrestrial phenomenon, with fish and whales in the oceans as a bonus. Actually the reverse is the case, but it's only with modern research techniques that the extraordinarily rich life of the oceans is beginning to manifest, with some 13,000 new species discovered in 2003 alone, out of a known total range of 38,000 species from plankton to whales. More than 90% of Earth's biomass (in weight of living matter) is found in the oceans, of which 90% is made up of single celled and microbial species, and 80% of ocean species depend on endangered coral reef environments.[10]

We think of the forests as the producers of oxygen, but marine plants, phytoplankton and algae convert much greater amounts of CO_2 into oxygen. Water vapour rises from the ocean surface to form clouds which release fresh water over the land. Ozone rising from the ocean's surface protects life from ultraviolet radiation. These sources of oxygen are under threat, just as are the forests.

Containing approximately 97% of the Earth's water, the average depth of the oceans is almost 4,000 metres. It is thought that life started with *extremophile* species around the hydrothermal vents on the deep sea floor that can heat the water up to 450°C. Similar rare creatures are still found today. They do not need photosynthesis for food, as does 97.7% of the biosphere. They obtain their energy solely from chemical reactions, and were the first step in the evolutionary process.

With the Sun, the oceans and the atmosphere are the principal drivers of the Earth's climate. They moderate it by removing excess heat from the Tropics through the thermohaline oceanic circulation system. Although phytoplankton comprise only 0.2% of the world's biomass, their prodigious reproduction rate helps account for nearly 50% of the Earth's primary food source, and support the incredibly rich marine community, from zooplankton to whales; but phytoplankton thrive only in cold waters. The rich biodiversity of the oceans is now under threat from decimation of the base of the food chain through global warming, and the higher part of the food chain by industrialized over-fishing and pollution.

The importance of water for evolution

Viktor Schauberger (1885–1958), the pioneering Austrian ecologist, saw water (including sap and blood) as an organism, the vital life-giving and energy-empowering vehicle that intelligent Nature uses for all forms of transmission and communication, energetic as well as physical, a medium linking Earth and Cosmos, with a vital role in promoting higher evolutionary life-forms.

Schauberger emphasized the importance of the water cycle. This intricate cycle, together with the minerals that it supplies, can be seen as providing the nourishment for evolutionary biodiversity. Water, as the vehicle for life, may indeed have its own evolutionary journey. Fresh water in Caledonian times may have been much less complex than today's — not in its chemical composition — but in its structure.

As we shall describe in later chapters, water is able to carry more intricate information when its structure is more complex. In Caledonian times, four hundred million years ago, to have a simple laminar structure was perhaps all that was required of water. The evolutionary needs of new life-forms would demand more complex DNA and chromosome forms. To support this, water would need to develop more complex geometric structures, for instance, three-dimensional geometry, complex octahedrons, molecular clusters, Platonic solids and their derivatives. These structures appear now only in the highest quality of water, indications being found in techniques like the drop water method of analysis (see Chapter 13).

The evolution of mammals was a giant step forward in evolutionary sophistication, which could not have arisen without the development of water structures more similar to the complexity we find today (see Chapter 3).

The civilizations of Man could not have evolved without mountains providing the minerals, the water and the fertile land. Ice and water erosion were the drivers of this vital process. Great alluvial flood plains became the cradles of civilization — Mesopotamia, the Indus Valley, the Yellow River, the Nile delta. Mountains provide the minerals for fertility and are the source of the great rivers. It is therefore not surprising that many people find mountains spiritually nourishing.

The skin of the Earth

The skin of any organism performs a number of important functions. As the outside layer, it defines the integrity and coherence of the organism and protects it from physical assault and infection. It is the vital heat-balancing organ for most organisms. It is also the antenna for receiving deeper energies, cosmic, or earth energies (see Chapter 13). The skin's health is important in effectively performing these functions.

The Earth's 'skin' is rather different, but the principles are much the same. It is composed of atmosphere, oceans, mountains, ice and life-supporting land or deserts. They work as a single system, for if one part of the system becomes unbalanced, the integrity of the whole is affected. Ninety per cent of the ice-free landmass in the current epoch was normally covered by vegetation. 75% of this had forest cover; this is now only 25%, causing modification of the world's climates, loss of fertility and soil erosion. On 'the planet of water', it's no surprise that the land skin is water-dependent. Its absence causes desertification.

The Moon and its tides

When you consider that the oceans are a thin fluid skin of the Earth, it is easy to see that the Moon's mass exerts a gravitational pull on them. The Sun does too, but being 390 times further away, its gravitational influence is much smaller. As it circles the Earth, the Moon pulls all the water bodies. So, besides the tides of the oceans, it affects lakes, rivers, air masses and rainfall; thunderstorms with their ionizing effects.

There are Earth tides as well; a Columbia University study carried out across America in 1970 found that the land surface rises and falls an average of 12 inches each day. And there are detectable lunar winds, in the morning flowing east, and west in the evening, affecting plants. All water-based organisms are affected too.

The Moon's orbit is elliptical; when it is closest to Earth (its perigee) the tidal effect is stronger. Its orbit is also tilted; when it crosses Earth's orbital path (the nodes) it can eclipse the Sun's light. Spring tides are caused by Sun and Moon being in the same line, which happens every fourteen days. The highest tides occur every 7 ½ lunations, when the perigee coincides with a New or Full Moon. If a storm happens to be moving onshore at this time, flooding and property damage can be severe.

Long estuaries or narrowing bays, like the Severn estuary on the west coast of England or the Bay of Fundy in Nova Scotia (Canada), have differences between high and low tides of as much as 40 feet. There are also nodal points where ocean currents meet, which can cause larger variations.

Inter-tidal environments are very productive eco-systems, often having the richest biodiversity, from rock pools, to coastal marshland and mangrove swamps. Inevitably they are rich in nutrients, so they attract a wide range of small marine animals, insects and birds. They were the stepping stone for life to emerge from the sea and make tentative land colonizations.

The Moon has a strong effect on many biological processes in organisms, such as reproductive cycles, plant growth cycles and on water in Earth's crust. Ancient peoples understood this and, regarding the Moon's influence as more personal than the Sun's, some cultures used the thirteen month Moon calendar for the year. As the Moon affects all growing things, a system of lunar-based horticulture has grown up, which we shall consider in Chapter 13.

First we shall look more closely at the nature of water, this most familiar, yet strange substance.

3. Characteristics of Water

It is clear that life on Earth depends on the unusual structure and anomalous nature of liquid water. Organisms consist mostly of liquid water. This water performs many functions and it can never be considered simply as an inert diluent. It transports, lubricates, reacts, stabilizes, signals, structures and partitions. The living world should be thought of as an equal partnership between the biological molecules and water.

Martin Chaplin[1]

After hydrogen, water is the most common molecule in the Universe. It is the most common substance on the Earth and is inseparable from biological molecules. Water vapour is the principal greenhouse gas and is also responsible for absorbing 70% of cosmic radiation.

As we shall see, life depends on the curious structure and anomalous nature of water.

Basic water

Many of the characteristics of water are well known. It might be useful though, to consider its main chemical and physical properties:

- —Under normal conditions, water has neither smell nor taste. The intrinsic colour of water and ice is a very light blue.
- —Water vapour is invisible; mist and cloud are a mixture of vapour and small liquid particles.
- —It is transparent, allowing light to penetrate for photosynthesis in aquatic plants.

—Water as H_2O is an electrically polar molecule: the electronegative oxygen atom ensures that the two hydrogen atoms are relatively positively charged. This charge difference is called a dipole, which makes the molecule polar and stable. (See Plate 3.)

—The strange anomalies of water derive from the unusual geometry of the water molecules, which allows tight three-dimensional bonding that produces a particularly strong and stable structure. This means that although more kinetic energy is required to change water's state — for instance, a higher temperature to boil it — the bonds can also easily come apart.

—These hydrogen bonds are the clue to water's behaviour; they assemble and pull apart millions of times a second, giving water its extraordinary adaptability: for instance, water can climb up a tube by capillary action, which is at the heart of biological life.

—The boiling point of water is related to barometric pressure: while at sea level water boils at 100°C *(212°F)*, at the top of Mount Everest it boils at 68°C *(154°F)*. In the intense pressure of deep ocean trenches, water is still liquid at very high temperatures.

—If water behaved normally (that is, like its chemically-related compounds), it would be a gas at ordinary temperatures and there would be no rain, rivers, vegetation or body fluids. It is the only compound that is found in Nature in all three basic states — solid, liquid and gas.

—Earth's climate is largely regulated by the ability of water to absorb and retain heat (called latent heat). This enables water vapour in the atmosphere to condense, form clouds and produce rain. The amount of water vapour that the atmosphere can hold increases with rising temperature.

—Water has the highest specific heat of vaporization (the amount of heat energy required to raise the temperature of a given amount to form vapour from the liquid state), of any compound other than ammonia. This consequence of the strong hydrogen bonding helps to control large fluctuations in temperature, thus moderating Earth's climate. It is also related to water's capacity to retain heat.

—Water has a high surface tension because of the strong bond between the atoms: it sticks to itself and to other surfaces, assisting its capillary action in plants. Surface tension also causes water to form naturally into drops.

—Spring water often has the highest surface tension. The strong skin of this water surface is called the meniscus. Damaging water's skin can affect its ability to form membranes within its body.

—Water's most important quality is as a powerful solvent. Substances that dissolve easily in water (hydrophilic, or 'water loving') are salts, sugars, acids, alkalis, and some gases: especially oxygen and carbon dioxide (carbonation). Those substances that do not mix well with water (hydrophobic or 'water fearing') are mainly oils and fats, which are non-polar molecules.

—The components for cell functioning (proteins, polysaccharides and DNA) are also dissolved in water.

—Pure water has low electrical conductivity (permittivity), but as soon as salts are absorbed, it becomes a good medium for bioelectrical storage and transmission — the reason why biological water and sea water are saline.[2]

—Water achieves its maximum density at 3.98°C *(39.16°F)*, expanding 9% on freezing. This causes ice to float, allowing life to flourish below the frozen surface.

—Water is responsible for aromas. Hydrophobic oils and volatiles were converted from water by alchemical processes.

Water's structure

We think of a substance's chemistry as a description of the particular chemical elements it contains (like H_2O). But the way that the atoms, electrons and nuclei are arranged (their shape and structure) is vital to the behaviour of the substance.

In water's case, the geometry contained in the molecule determines what is unusual about it. Many chemists say that it is precisely the asymmetry of its molecule that allows water to perform its life-creation role.

The water molecule is a simple union of two small hydrogen atoms with a positive electrical charge, and one relatively large, negatively charged oxygen atom, arranged at an angle of 104°. (See Plate 3.) The oxygen atom has two very negative electrons, which drive the hydrogen atoms closer together, making the valence angle tighter. This geometry makes for an inherent instability to which most scientists ascribe water's extraordinary adaptability in a world made up of imperfect structures that is nevertheless programmed towards a state of equilibrium, harmony and beauty. As Paolo Consigli points out:

> The imperfect symmetry of the water molecule, with its bifurcated disposition, holds the secret of matter's very existence: instability and imperfect symmetry belong not only to the aquatic sphere, they are the guiding law of the Universe since its inception.[3]

Oxygen's negative charge holds the much smaller hydrogen atoms in close embrace, which makes for a very strong covalent bond that is easily made, but hard to break.[4]

We normally associate the term 'structure' with solid matter in which the stable matrices of atomic bonds keep their shape and form. Water is a medium that is constantly changing; how can it have a structure? Certainly, it seems that ordinary water cannot have a long-term order, but the key to its structure lies in the fairly stable non-crystalline clusters.

One of the most exciting areas of biological research is how its structures influence the role of biological water (in the body). Most chemists identify a tetrahedral (four-sided) structure to the water molecule. Martin Chaplin of London's South Bank University, who specializes in water's role in health, has proposed for the structure of water an ingenious model of a 280-molecule icosahedron water cluster, which is a highly symmetrical and aesthetically pleasing structure. It is possible, with this model, to map all the anomalous properties of water. (See Plate 28.)

The icosahedron is a twenty-faced polygon and the most complex of the five Platonic Solids which, in classical times, were thought to represent the building blocks of all of life and of the Universe.

The icosahedron was identified with the water domain, and with the qualities of personal transformation and sexuality, as well as the emotions.

Two states of water

The great mystery is: how is it possible for water to behave quite differently from ordinary liquids? The icosahedral model facilitates the mapping of the controversial idea, first proposed in 1901 by Wilhelm Röntgen (who discovered X-rays) that water exists in two states: bulky, low density (super-cooled water), and high density, where the molecules are packed more closely together.

The sixty-seven anomalies that water exhibits, listed by Martin Chaplin (see Appendix), can be explained by the two-states of water: for instance, liquid water is denser than in its solid phase, compared to other liquids which become denser in their solid phase, as their molecules pack closer together. Liquid water can be cooled below its freezing point without becoming ice, but when it is heated, it shrinks instead of expanding. When under pressure the maximum density point lowers, whereas with ordinary liquids it is raised.

Water as a solvent

The simplicity of the water molecule makes its electromagnetic qualities more effective. The positive charge of the hydrogen atoms attracts negative ions from the substance with which it is in contact. Oxygen's double negative charge in turn attracts the positive atoms of the substance. So, the combination of the positive and negative charges will start to break down the molecular structure of the wetted substance.

In this way, water breaks down and dissolves substances into its constituent parts. From the atmosphere it takes nitrogen, oxygen and CO_2, while from rocks it absorbs potassium, sodium, calcium etc. It is constantly moving around the building blocks of life, taking from one source and depositing them elsewhere for new growth, in a manner that appears to be not entirely accidental.

Water carries, as solvents or solutes, substances which can be damaging to life, such as heavy metals or chemicals from household, agricultural and industrial transfers which can be substantially removed by physical filters. Nature has her own way of dealing with them.

Water conveys energy

Water's main biological role is to carry information and transfer energy — beneficial when balancing an organism's energy. It also retains the detrimental energy of toxic substances, which can be as damaging to human health as the physical substances themselves.

Our modern lifestyles pollute many water supplies. Rivers, lakes and underground water are frequently polluted with metals and chemicals. It is inadequate to 'purify' these by filtering or adding chemical disinfectant.

Wolfgang Ludwig, an internationally regarded microbiologist from the Technical University in Munich warns:

> After the water treatment has 'purified' the water as far as the 'science of yesterday' is concerned, it still carries certain electrical frequencies, oscillations in specific wavelengths. By further analyses, these can be tracked precisely to those detrimental substances which were physically detected in the water before treatment.
>
> Certain electromagnetic frequencies of heavy-metal polluted water have been found in cancer tissue as well. Let us take the very low frequency of 1.8 Hertz. We have been able to confirm that fresh water in a certain major German city carries this 1.8 Hertz frequency, even though that water was distilled twice before this measurement!

Water and life

Water is an essential part of the two metabolic processes in the body: *anabolism* — in condensation reactions the elements of water are yielded from the combining of smaller molecules to *make* larger ones (for instance, starches and proteins) for fuel storage, structural components and information; and *catabolism* — water is used to *break* bonds to make smaller molecules (glucose, fatty and amino acids); for instance, to produce energy.

Water is essential to photosynthesis, the basic energy-generating and building process of life through which plants use the Sun's energy to separate water's hydrogen from oxygen. Hydrogen is then combined with CO_2 from air and water to produce simple sugars for the plant's growth and release oxygen to nourish higher life-forms.

An anomalous substance

Compared to any other known substances, water is pretty weird — it doesn't behave as it should, or really like any other liquid. The fact that ice floats, for example, not only means that life can flourish below the frozen surface of a water body, but also allows reflection of sunlight from its surface. Most substances are much denser in their solid than in their liquid form. The molecules of other liquids get packed tighter together on freezing, but with water the molecules move apart. Because of the hydrogen bonding, the freezing point of water is much higher than expected, but the transition from liquid to solid is easier because ice has 15% fewer hydrogen bonds than in the liquid state.

Other hydrogen bonded liquids in the hydrides family have boiling points that relate to their molecular weight. If water did, it would boil at –80°C instead of 100°C. Other solvents and alcohols boil at much lower temperatures, between 38° and 80°C. The very high boiling temperature for water is because of the very extensive hydrogen bonding of the molecules. It takes a lot more energy to break apart those tight bonds (see Figure 2).

The main reason that the Celsius temperature scale is based on the freezing and boiling points of water is because of its familiarity, not its normality. It is the same with the measurement of specific heat (the ability to store heat), where water again is the benchmark. Once heated, it takes a long time to cool off. Water has the benchmark specific heat of 1 (1 calorie of heat needed to raise 1 gram of water by 1°C). This is high compared to ice, which is 0.5, and to water vapour which is

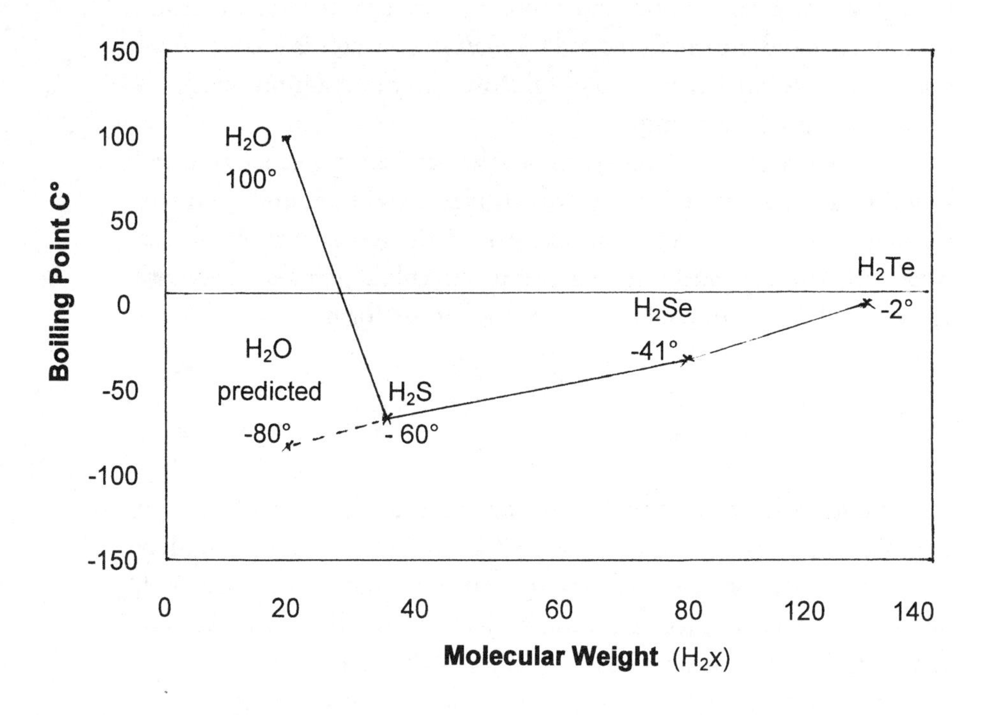

Figure 2. Hydrides boiling points. Members of the family of hydrogen compounds have boiling points related to their molecular weight — except for H_2O. This particular anomaly is crucial for making water fit for the maintenance of life. (Jill Granger)

0.47.[5] Mammals could not survive without the heat storage capacity of their bodily fluids. Water is a very efficient heat storage medium for buildings, too.

Water is much more responsive to changes in temperature and pressure than are other liquids. From 4°C, when at its densest, water expands on cooling *or* heating. When cold water is heated, it shrinks; when hot water is further heated, it expands. With increasing pressure, hot water molecules move slower, but cold molecules move faster. It would not be an overstatement to say that life depends on the weirdness of water. Its high specific heat keeps the body warm in the cold, and the high latent heat of evaporation (that is, a lot of energy is needed to change water into a vapour) allows the body temperature to stay within healthy limits. A dog may not sweat like a human, but when he's hot, just watch his tongue and how he expels water from his body!

In a large body of water, density-driven convection takes life-giving oxygen to the depths. The large heat capacity of the oceans means they can act as heat reservoirs, fluctuating far less than land, being cooler in the summer and warmer in winter, thus moderating the world's temperature. We see that a coastal environment always has a more moderate climate than inland. Because water releases a lot of heat before it freezes, this has the effect of moving the planet's coldest zones towards the poles, extending the temperate habitable zone. The Gulf Stream's warm water gives Western Europe a much more pleasant climate than it would have otherwise.

The 'intelligence' of water

Moving from the more conventional attributes of water to the more unorthodox, we use the term 'intelligence' in connection with water because it optimizes the conditions for life by regulating environmental processes. At a temperature of 37°C *(98.4°F),* water requires the least kinetic energy input in order for its temperature to rise, allowing it to keep the blood in the human body at a constant temperature of 37°C, any deviation from which indicates illness.

Another of the 'intelligences' of water is that it takes a lot of energy to freeze and, once frozen, to melt. A layer of ice formed on a water body's frozen surface slows down the deeper process of freezing. A large amount of heat is extracted from the water to allow it to freeze, which warms the environment and slows down the freezing process. Similarly, because water extracts huge amounts of heat from the air on evaporation (which is why sweating is so effective), the hottest zones are pushed back towards the equator.

The third intelligence of water is due to another of its anomalies — the huge amount of energy required to vaporize water (called the latent heat of evaporation). This stops the air from excessive heating, as water retains its heat before evaporation. If the air is cold, the condensing water releases heat into the air. This is a very effective balancer of climatic extremes. The warmer the air, the more water vapour it can hold, and the longer the temperature can remain constant.

The interaction of these three properties of water has created a temperate climatic zone on Earth with optimum conditions for evolution of the human species. Certainly complex civilizations could not have evolved without the stability of moderate, stable climates.

This variety and complexity of regulatory systems are so extraordinary that it is hard to believe that the manner in which they developed in stimulating evolution was purely accidental. Without these self-regulating systems, life could never have developed beyond the simplest forms.[6]

One of the most important functions of water is to facilitate cellular functions in the body. Because of its unusual hydrogen bonding, it has the unique ability through hydration to activate proteins, and through its ionizing abilities to facilitate proton exchange and cell formation. Its particular ability to act as a solvent is essential in the action of salts and ionic compounds.

The degree to which water is the ideal substance to make living processes succeed begs the question: 'Did the unique properties of water facilitate evolutionary development in organisms?' Neo-Darwinists proffer one explanation of this. In Chapter 9 we shall examine a different possibility.

The anomaly point of water

The density of water is crucial to its behavior. It is at its densest and has its greatest energy content at a temperature of 4°C *(39°F)*. This is the so-called 'anomaly point', which has a major influence on its quality. Viktor Schauberger called this temperature 'the state of indifference' for water, meaning that when in its highest natural condition of health, vitality and life-giving potential, water is at an internal state of energetic equilibrium and in a thermally and spatially neutral condition.

It was very convenient of Nature to arrange that mammals and other creatures should depend on blood for homeothermic balance. In the body, the temperature of the blood (composed 90% of water) is almost exactly the same as that of water at its most stable temperature of 37°C *(98·4°F)*. This means that our bodies are able to tolerate a wide range of ambient temperatures, for a great amount of heat or cold is required to change the temperature of water. But it also holds on to heat well; good for both body temperature and for domestic heating systems.

Coherence

Water appears to be a disorganized medium. The order created by its hydrogen bonding networks seems to have a very local effect. In order to understand how water is able to display such integrative roles in organisms, scientists have for decades been using X-ray bombardment to look, unsuccessfully, for evidence of a more long-range order.

This principle may be similar to that of water clumping. Recently two chemists in South Korea discovered that, contrary to the laws of chemistry, when a solution is diluted, the water solutes bunch (clump) together, giving the water more coherence. This may be a key to how homeopathy works.[7] (See Chapters 11 and 15.)

Recent research using Magnetic Resonance Imaging has shown how salt dissolved in water quickly spreads throughout the whole water body. Dr Masaru Emoto has used a similar imaging technique in his research with ice crystals (see Chapter 15).

It appears that this long-range order or coherence is a quantum effect. The problem for mainstream physics and chemistry is that the quantum domain has different laws from the classical Newtonian. Newtonian physics-trained biochemists struggle to understand the molecular behavior of water. Quantum physics traditionally studied the sub-microscopic world. However Dr Mae-Wan Ho, following recent research by Frohlich and Popp has found that there are quantum coherence qualities in the organism, which are now being considered by a number of physicists.[8] We shall look further at this research in Chapter 11.

Viktor Schauberger, an unorthodox researcher, established a wonderful picture of the behaviour of water. His peers mocked him when he insisted that water behaves like a living organism. (Wouldn't they still today?) When it reaches maturity water displays amazing properties. He showed how, when it is vibrant and healthy, it pulsates, twists and spirals in a very specific way that maintains its vitality and purity, enabling it to fulfil its function for all organisms as an energy channel and a conveyor of nutrients and waste.

It is important to identify the difference between orthodox and unconventional ideas about energy. The orthodox sees energy in water as the ability to do work, as in the potential of a column or water to drive a turbine. The unconventional picture is of an electromagnetic energy to do with polarities, resonance and the transfer of information from one state or order to another; we call these subtle or dynamic energies.

Schauberger said that, when it is immature, water takes, absorbing minerals with a voracious appetite, to give back the much-needed nourishment to its environment only when mature as in an established river. Water has a memory; when we think we have 'purified' our water supplies of the contaminating chemicals and hormones, their subtle energy remains, as well as that of the chemicals added as disinfectant to make it 'drinkable'. The subtle energy of these contaminants remains, polluting our energy bodies in the same way that chemicals affect our physical bodies. Because of its nature, water sacrifices itself entirely to the environment, for good or for bad.

We can observe water's coherence in everyday situations. If we watch water streaming down an inclining road after a shower of rain, or as a rivulet on the sloping beach sand towards the sea, we will notice how it pushes down in a jerky rhythm, as pulsations. That is because water is 'alive' — it is inherent to its nature that it actually does pulsate, just as blood pulsates through the veins and arteries of the body. But the most extra-ordinary fact about water that Schauberger discovered is that it has the power of self-purification, and can restore its generative properties in the same way that other living things can heal themselves.

Properties of water

We shall note different properties of water in the appropriate themes of subsequent chapters. Suffice for now to note in general that there are properties connected with flowing water, like turbulence, which may be either totally chaotic water behaviour that can be destructive, or it may be purposeful chaos produced by a rock in a stream which can create a train of longitudinal vortices that introduce higher energy into the stream body.

Other properties may be turbidity, found in mature streams to stop overheating of the water body and keep the bottom water more active. Viscosity relates to the ability of water to flow — its degree of 'thickness'. Water of higher viscosity is found near springs (when the meniscus or 'skin' is better formed) and lower viscosity water is more 'wetting' and more easily absorbed. Further properties that are taken into account by conventional hydrologists when making their calculations, are tractive or shear (sweeping) force, flow velocity, and sediment load.

One can safely say that the properties of water are weird, looked at from both an orthodox and an unconventional perspective.

Ice

Ice — easier to study, having a stable structure — can be any of fifteen solid phases of water, depending on pressure and temperature.[9] The common one that we come in contact with is ice called Ih, which is crystalline and either bluish-white or transparent. When liquid water or condensed water vapour is cooled below 0°C *(32°F)* it forms ice. It can also form straight from water vapour, as frost forming on any solid object, or as snowflakes or hail, forming around a minute dust particle. It begins as snow crystals which, as they collect on a frozen surface, gradually compact into ice globules

We know so little about water as yet. It is likely that, in time, at least as many types of water will be identified as there have been of ice.

We shall now observe how the water of the Earth circulates through its rocks, sloping hills and deep underground.

> *Water is H_2O, hydrogen two parts, oxygen one, but there is also a third thing that makes it water and nobody knows what it is.*
>
> D.H. Lawrence (1885–1930)

4. The Blood of the Earth

Were water actually what hydrologists deem it to be — a chemically inert substance — then a long time ago there would already have been no water and no life on this Earth. I regard water as the blood of the Earth. Its internal process, while not identical to that of our blood, is nonetheless very similar. It is this process that gives water its movement.

Viktor Schauberger[1]

Water and the health of the Earth

Water in the Earth works like blood in the human body; it performs similar roles — to nourish, communicate and recycle. Likewise, the health of the Earth's water is as much an indication of the health of the whole organism as is blood to the body, as Attenborough and Schauberger have both warned.

We do not treat water as the most precious substance for life on Earth because our current scientific model does not acknowledge water's extraordinary properties. This critical situation will change only when a new generation of scientists can recognize the Earth as an organism and humanity as truly part of Nature, subject to her laws. This is the first priority, even before we consider what 'treatment' is necessary.

The Earth's water domain is going through enormous changes brought about by global warming. Climate change is altering the behaviour and distribution of rainfall, exacerbating extreme weather and melting the glaciers which have been a valuable storehouse of fresh water and maintenance of river flow for the millennia of humanity's existence on the Earth. The oceans are losing their ability to store CO_2 and their fauna is suffering stress, causing migration and species loss.

The water planet

Earth is known as the water planet. The wonderful, blue colour it presents from space derives from the abundant amount of water in its atmosphere (which optically screens out the red end of the spectrum) and from its surface being 70% covered by the oceans. The atmosphere is composed 78% of nitrogen and 21% of oxygen. The greenhouse gases, which act like a blanket preventing the planet from losing too much heat, are a tiny fraction of the total atmosphere, spatially limited mostly to the layer known as the troposphere. Water vapour accounts for 60% of the greenhouse effect, CO_2 for roughly 36%.

As the principal component of this planet's surface, water is constantly moving, circulating and interchanging. In its vaporous state it swirls in great streams, moving heat and cold from one latitude to another, balancing the temperature. Humidity, wind and temperature, and their changing behaviour are the basic ingredients of atmospheric weather.

Despite its ubiquitous presence, water prefers to play a mysterious role. Like the chorus in a Greek play, it stays in the background, yet it is the foundation of all processes. Its role is a quandary, on the one hand passive — for it does not in itself initiate processes, being the vehicle for change — on the other active and facilitating. There is no question but that water rules the world. The outcome of energetic processes may be beneficial or catastrophic but water, in one of its three states, makes it all happen.

Our incomplete knowledge of water's role and function makes it hard to predict the processes of instability and change that we are experiencing today. Water magnifies and accelerates a process like that of climate change. The future of humanity on the planet will to some extent rest on our ability to understand that the critical points of the warming cycle are determined by positive feedback loops that depend on water. For example, we have only recently come to understand that the oceans are the largest store of CO_2.

The vast oceans, teeming with life, have stored enormous amounts of minerals and salts washed down from the continents over three billion years. They have their own complex circulatory system, which balances temperatures and nutrients. The land masses receive most of their fresh water as precipitation derived from the oceans. Much of this finds its way into the surface of the Earth, forming artesian basins and enormous inaccessible aquifers. But the Earth also has even greater amounts of ancient water locked up in rocks and volcanic material from its very formation, and may also be making new primary water. All of this water circulates, carrying energy.

The habitable mass of the oceans is hundreds of times greater than that of the land and the atmosphere just above it. Although it has been calculated that 90% of life is in the oceans, we still know much less about the oceans and their history than about the land. However, we shall be more concerned in this study with fresh water. Water, as the vehicle for life, may indeed have its own evolutionary journey, as part of its parent Earth's history.

Distribution of water

Ninety-seven percent of the world's water is in the oceans. The remaining three per cent is fresh water, of which:

Ice sheets and glaciers	75% (much of which will be lost by global warming)
Very deep ground water	14% (inaccessible)
Less deep ground water	10% (for instance, aquifers)
Lakes	0.3%
Rivers	0.3%
Soils	0.06%
Atmosphere	0.035%

Ice sheets and glaciers

For the last three thousand million years, Earth has switched to and fro between a world that basks in a greenhouse climate (90% of its existence), and one that is largely covered by ice. There have been at least four major ice ages. The last one started about 1.8 million years ago, and has still not ended, with vast ice sheets still covering Greenland and Antarctica. The ice started as snowflakes which consolidate into grains of ice with the weight of continually consolidating snow.

Earlier in the current ice age, continental sized ice sheets spread over North America and North-west Europe. Icecaps and glaciers ground down the bedrock, carving out huge valleys. Ice has had more impact on man in the last two million years than has any other environmental influence, principally because of its beneficial climatic effects.

Man has evolved during the Pleistocene Ice Age, which had some four major advances, interspersed with warm interglacial periods when the climate was sometimes warmer than now. In times of warming, glacial meltwater formed huge lakes that, when the dams that held them back were eroded, caused inundations on a continental scale.

Earth has some 30 million km^3 of ice, enough to cover the entire planet to an average depth of 60 metres *(200 feet).*[2] Ten per cent of Earth's landmass is snow and ice (84.16% in Antarctica, 13.9% in Greenland, 0.77% Himalayas, 0.51% North America, 0.15% South America, 0.06% Europe.[3] However, scientists are now predicting that enough of this will melt by the end of this century to flood most coastal cities and plains bordering the ocean.

The Greenland ice cap has been experiencing unprecedented and accelerating melting in the last decade; the whole structure is becoming unstable. It has enough to raise sea level by 6 metres if it all melted. Its rate of melting is accelerating, but still occupies 80% of the island up to a depth of 3,000 km and a volume of 3 million km^3. There is a risk that the freshwater melt flowing down either side of Cape Farewell (Greenland's southern point) could cause the beneficial Gulf Stream's motion to stop (see p. 91).

In the Arctic what has surprised scientists is the speed of melting of the sea ice. At the present rate there might be no summer sea ice by 2020. It has a strong albedo effect, reflecting the summer sun to keep the arctic waters cool. When this is lost the summer warming will accelerate.

The Antarctic icecap covers 13 million km^2, one and a half times the area of the USA's fifty states, and contains 30 million km^3 (7.2 million cubic miles) of ice. It has an average thickness of approximately 2,000 metres (6,600 ft), in places reaching depths of 4,000 metres (13,000ft) or more. The Antarctic summer (when the ice melts) has lengthened from sixty to ninety days since the 1970s. Since measurements began in the 1950s, the average temperature on the Antarctic Peninsula has risen by approximately 2.5°C *(4.5°F).*[4]

The most dramatic sign of change in the Antarctic in 2002 was the sudden breaking off of the Ross Ice Shelf, 400 feet thick and the size of Cornwall. As it had been floating on the sea, its melting has not much affected sea level.

Polar ice, through its 'icebox' effect, makes the temperate latitudes much more habitable than when the Earth is unglaciated. This cooling is amplified by the negative feedback of the white surface reflecting the Sun's energy in summer. The world's atmospheric and oceanic circulatory system balances the climatic contrast between the tropics and the polar region, creating a climate fit for a complex species like Man to develop his enormous potential (for bad and for good!).

An unglaciated planet, without polar icecaps and a developed system of mountain glaciers, does not favour Man's evolutionary potential. It is becoming clear that, without the polar ice to balance tropical heat, the Earth's climates will become more extreme. This changed environment will be hostile to our species and will almost certainly contribute to a rapid fall in the current world population level.

The glaciers of the Himalayas and associated ranges are the 'Third Pole'. They feed the giant rivers of Asia and support half of humanity. The annual melt from the world's mountain glaciers has been a valuable source of fresh water in hot summers. Its disappearance will cause much deprivation for the people dependent on this, particularly in China, India and central Asia. (see Chapter 17).

If the glacial meltwater disappears from the lush and productive Kashmir valley (reclaimed from a primordial lake) there will be no summer surface water in the valley. If the groundwater is not replenished by the melt, it will no longer be known as 'Paradise', but will become desiccated and barren.

We live in transitional times. Theoretically we are still in one of those exceptional periods favoured by Nature for evolutionary accelerations, an ice age. Yet, clearly the environment is changing before our eyes, and the world will look very different by the end of this century. Still, the great water cycles will continue, with increasing potency.

Deep groundwater

There is an enormous amount of water distributed through the deep crust of the Earth. An aquifer is formed where it collects in a basin, absorbed by layers of permeable rock or sand. The upper limit may be the water table, or the layers may be many hundreds of metres below the Earth's surface and millions of years old. American, Australian and Russian scientists have discovered substantial water resources in the Earth's mantle. One such is a body equal in volume to the Arctic Ocean beneath eastern Asia at a depth of roughly 1000 km (620 miles).

In total, these underground resources probably represent 90% of the fresh water on the Earth. Only a small amount of this is accessible, and historically any drawing off was always replenished by rainfall seepage. In the last century, industrial farming techniques and the growth of the cities have created enormous demands for water which could be supplied only by exploiting the accessible aquifers, far beyond their ability to be replenished.[5] As much as 80% of all fresh water is consumed by irrigation, much of which now comes unsustainably from aquifers.

There is, as yet, little understanding of how to gauge when extraction of resources from an aquifer is sustainable — that is, that the refill balances the extraction. In fact most of the world's aquifers are fast depleting.

The quality of the water varies considerably. The vast Artesian Basin of Australia has a high sodium content, making it unsuitable for crop irrigation. As a powerful solvent, water will pick up and dissolve many chemicals and gases, such as sulphates and sodium or the radioactive gas radon. In India, serious groundwater contamination by arsenic and fluorides has led to bone deformities and crippling organic damage.

Lakes

Lakes contain the largest amount of accessible fresh water. Many are found in recently glaciated terrain where natural dams were formed by retreating ice sheets or glaciers (for instance, the Great Lakes of USA/ Canada). The very deep ones are found in regions of tectonic movement or in rift valleys (Lake Baikal in Siberia, the African Rift Valley lakes,

Loch Ness in Scotland); they are the oldest, and are often still growing in depth and volume. Some form in volcanic craters (Crater Lake, Oregon), or under an ice sheet (Lake Vostok under Antarctica, which is enormous). Then there are salt lakes, which have no outlet (Caspian and Dead Seas).

Deep lakes tend to be layered so that the cold lower water is not disturbed. Nutrient-rich lakes have a rich fauna and flora, sometimes with algal blooms and poor ecosystems due to lack of dissolved oxygen.

Many lakes are now severely polluted if they are fed by rivers from agricultural land or are near cities. Lake Baikal, the world's oldest, largest (in volume) and deepest lake contains about 1,700 species of plants and animals unique to the lake. It holds 27% of the world's surface fresh water, of pristine quality. This is now under threat from a planned uranium processing plant in the vicinity. Environmental protesters have been harassed by the Russian government.[6]

The river is everywhere at once, at the source and at the mouth, at the waterfall, at the ferry, at the rapids, in the sea, in the mountains, everywhere at once ... there is only the present time for it, not the shadow of the past, nor the shadow of the future.

Hermann Hesse, *Siddhartha*

5. The Great Water Cycles

The atmosphere is ... not merely a biological product, but more probably a biological construction: not living, but, like cat's fur, an extension of a living system designed to maintain a chosen environment.
James Lovelock, *Gaia*

Water is the engine and the workhorse of the planet. It is always moving, and doing something. The nature of these movements is cyclical. There are great cycles and little ones; slow and fast ones. Water is always circulating, an exchange between the deep, steam-heated water in touch with Earth's magma, and the deep groundwater above the mantle. In the slow cycle, groundwater is held in deep aquifers or glaciers for millennia, gradually seeping through to the surface through springs or from melting glaciers. The fast cycle lubricates the active parts of the atmosphere, with 500,000 km3 of water evaporating annually from the ocean and land as vapour, condensing into clouds and precipitating back to the surface. There are also the cycles of the circulation systems in organisms.

The great aerial ocean

The Great Aerial Ocean is a term coined by Alfred Russel Wallace, co-founder with Charles Darwin of the Theory of Evolution by Natural Selection. It well describes the intimate relationship between the oceans and the Earth's atmosphere, the Sun driving the great water cycle. The heat received on the Earth's surface is greatest at the equator, which leads to a transfer of heat from the equator towards the poles. Three quarters of this heat is carried by water in the atmosphere, one quarter in the ocean currents. This cycle is a fundamental role of water for the maintenance of life on this planet. The primary cycle is atmospheric; the atmosphere is warmed from below, while the oceans are heated at the surface. The surface winds drive the ocean currents, assisted by the spinning of the Earth.

The atmosphere

The atmosphere is, if you like, the front line of the water story. In its early days, Earth's atmosphere must have been very dark, hot and thick, with a high percentage of CO_2, hydrogen, sulphides and a lot of methane. This hostile mixture persisted for about two billion years. Then a remarkable thing happened. Cushion-formations of bacteria called stromatholites appeared — the first oxygen producers — which over the next billion years gradually built up enough oxygen in the atmosphere to prepare for the emergence of land-based animals, aided later by the photosynthesis of plants.

The atmosphere, for most of Earth's history, has been an unstable environment, disturbed by volcanic eruptions, cosmic collisions and other unpredictable events. It was only when the Earth calmed down sufficiently for life to become established on land that Gaia was able to develop climates necessary for the progress of evolution.

The geological record shows that there were times when climatic conditions were more balanced, more or less defined by geological periods, interrupted by violent earth changes typified by tectonic shifts, vulcanism, mountain building, cosmic collisions, global warming and cooling, ice ages, species extinctions and later by carbon dioxide shifts.

We don't know a great deal about the composition of the atmosphere in past geological eras. For the last 500 million years it must have been a mixture of simple gases, not dissimilar to ours today. The atmosphere is highly compressible, so that at 18,000 feet above the sea, it is only half as dense as at sea level. 90% of all the atmospheric gases are found below 15 km, yet the atmosphere extends up to 100 km above Earth's surface. The overall thickness of the atmosphere at the poles is half what it is at the equator. The Earth's gravity is what prevents the thin atmosphere from escaping.

The atmosphere is now composed 78% of nitrogen, 20.9% of oxygen and .08% of argon. The remaining .03% are trace gases. One is ozone, present in tiny amounts, which keeps us from going blind, getting sun-caused cancer, etc. Then there are the greenhouse gases, of which water vapour comprises 60%, and most of the rest is CO_2, with a smattering of methane and CFCs (chlorofluorocarbons).

The amount of water in the atmosphere varies considerably, depending on the temperature (cold air can hold very little), but doesn't exceed 4% by volume. Nevertheless, it is the water vapour that carries the heat and creates the weather, influenced by the amount of water carried (degree of humidity) and, of course, the temperature. Water vapour has a complex role in the atmosphere. We tend to think of CO_2 as the principal greenhouse gas, but water vapour absorbs about 70% of solar radiation, mainly in the stratosphere. The two gases act together creating positive feedback loops. So a CO_2 concentration warming the atmosphere allows it to take up more moisture, which then further heats the atmosphere.

Without plants and algae to soak up our waste CO_2, we would soon run out of oxygen and suffocate in CO_2. This is a self-sustaining cycle that is the foundation for life on Earth. The amount of carbon in the cycle is prodigious. Living things account for a trillion tons, while the amount stored in the ground is several thousandfold greater, and the oceans contain over fifty times what is found in the atmosphere.[1]

The atmospheric layers

Water vapour represents 75% of the total gas mass of the atmosphere. Temperature decreases with altitude up to the upper boundary, which is called the tropopause. This varies in altitude from 10 km *(6 miles)* at the equator to 8 km *(4 1/2 miles)* at the poles.

The atmosphere can be divided into several well-marked layers (see Figure 3). The lowest layer, the troposphere, is the zone of weather phenomena and turbulence. Despite its small volume, you could say that water controls the lower atmosphere.

The second major division is the stratosphere, which extends from the tropopause to the stratopause at about 50 km (30 miles). Ozone, the gas that protects life from ultraviolet radiation, is formed at the stratopause, and accumulates lower in the stratosphere. Modern jet aircraft fly in this level of the atmosphere. Above this is the mesosphere, which has sufficient gas particles to burn up meteors,

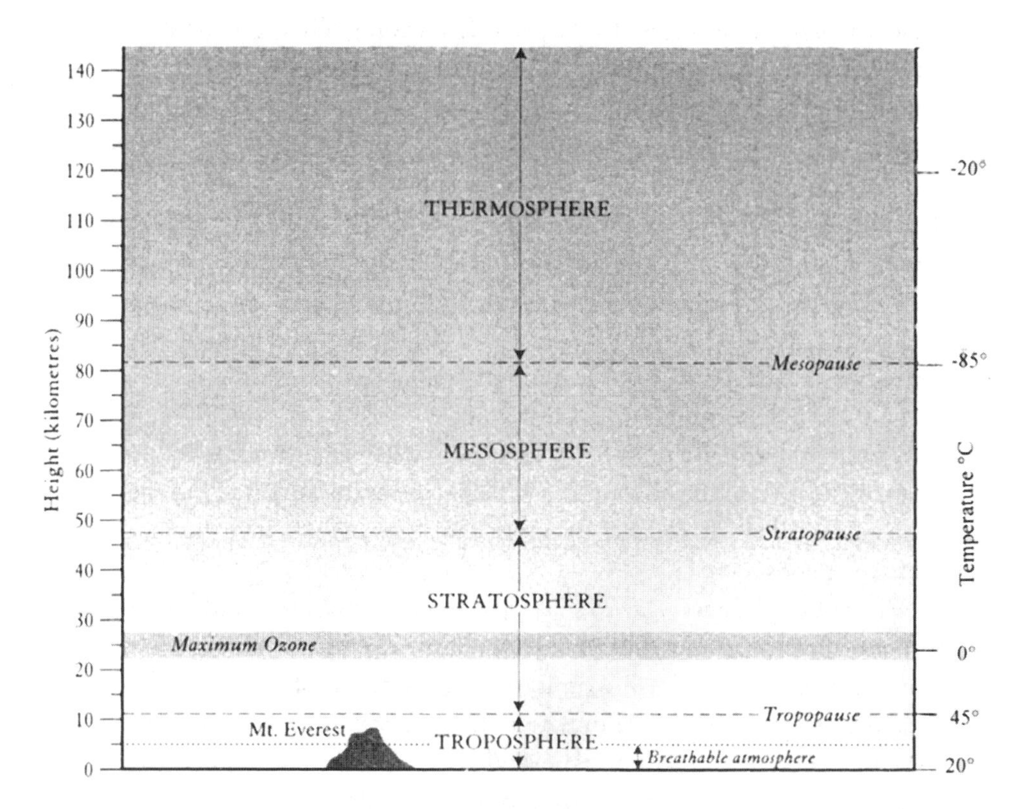

Figure 3. Cross-section of the atmosphere. The four major bands of the atmosphere. The thermosphere is also known as the ionosphere. There is a tiny amount (0.000001%) of ozone gas, without which we would go blind and get cancer. Only a small amount is breathable air — at 25°C (77°F) water vapour makes up 3% of what we inhale. Noctilucent clouds form at about 45 km height, and the aurorae borealis are found between 60 and 100 km. (Tim Flannery)

acting as a protection for Earth life. There is just enough water here for noctilucent clouds to form. Above this is the thermosphere at 85 km, the environment for the space shuttle and for the aurora borealis. This awe-inspiring phenomenon is thought to be caused by plasma particles from the Sun being deflected towards the magnetic poles and colliding with air and water vapour molecules, surrendering their energy as photon emissions[2] (Plate 18).

Climates

Earth's atmosphere is vital to terrestrial life. Its present form dates from probably Caledonian times about 400 million years ago, when the first forests and land animals became established, and high quality water was an evolutionary requirement. Before this, climates were perhaps irrelevant. The atmosphere became an invaluable shield against harmful radiation and, with vegetation, was the indispensable thermostat to ensure the optimum environmental conditions for evolution to proceed.

There is insufficient awareness of how much of the Sun's energy is reflected back to space by ice and snow (the albedo effect). The resultant cooling of the Polar regions produces a larger climatic contrast between the tropics and the poles than in unglaciated times. The world's atmospheric and oceanic circulatory system balances these out, creating a climate fit for a complex species like man to develop his enormous potential. The average world temperature is lower in glaciated times. As we have already noted above, an unglaciated planet would not favour man's evolutionary potential, unless the Sun's energy continues to wane in the future.[3]

A climate derives from the specific way in which the given energy from the Sun is affected by the local influences of humid ocean, a continental mass, temperature and topography. It will vary according to latitude and season. So, for example, the plains of India warm up in the summer, drawing in the damp maritime air from the tropical oceans. The mass of the Himalayas forces this damp air to rise and release deluges known as the monsoon.

Continental masses tend to produce areas of high pressure. Climates are the result of many local factors such as wind belts (the jet stream, the roaring forties, trade winds of the sub-equatorial region, the El Niño, and so on). A climate is also affected by the amount of turbulence in the weather. So, in the temperate latitudes, where there is more mixing of different types of air masses (see Ferrel circulatory system below) changeable weather patterns determine climatic variations. This is particularly true of a coastal environment like that of the British Isles. The El Niño effect, when the southern Pacific Ocean warm current changes its direction of flow, is at a maximum early in the 2010s, which is likely to have adverse effects on worldwide weather.

Air masses play a large part in both climate and the resultant weather. An air mass formed over the ocean will have a high humidity, while a continental air mass will have a low humidity. The boundaries between air masses are called fronts, which tend to be the breeding grounds for storms and unsettled weather. People respond very differently to varying humidity, to the amount of sun and to stormy conditions. The change in the seasons also affects people differently. This will often determine where one decides to work or to settle.

The British Isles are interesting in that they don't have a climate of their own, being subjected unpredictably to seven different climatic types. No wonder the weather forecasters have a difficult time! The islands are at the crossroads of several air masses, demonstrating how the huge differences in moisture content and temperature affect the resulting weather. The west side of Britain is influenced by the humidity of the ocean, and by the warming Gulf Stream; the east comes under the influence of the continental climates and northern air streams can have a great cooling effect.

There is also the powerful Jet Stream which flows like a giant river at about 120 mph. at altitudes of about 30,000 feet that commercial aircraft like to ride when flying eastwards. This can change its normal position from the ocean off Iceland to the south or south-east of Britain, and can have a disruptive effect on British weather, as it did in the summers of 2007 and 2008, when some areas experienced an average month's rainfall in a couple of hours and torrid heat baked the Black Sea and the Mediterranean regions.

Land heats up more than water but loses its heat more quickly (for instance, at night). Low pressure cells or depressions often come in a train, bred by weather fronts or perturbations in the major wind streams.

Weather and clouds

We usually think of weather as variety. When I lived in the USA where in summer the skies are often cloudless, I used to yearn for the fluffy clouds of England.

When we learn how special and magical is water, we might accept its gifts more graciously. Have you ever gone out in a light drizzle and tilted your head to let your face become refreshingly moistened? To walk barefoot on a dew-kissed lawn at dawn can be a most invigorating experience!

In winter, the weather can bring some astonishing effects, like the magic of a fresh snowfall which produces a strange silence in the landscape and muffled sounds. Or ice forming on tree branches or power and telephone cables.[4] The most magical of all is when fog freezes on a spider's web.

The atmosphere is largely invisible — except for clouds, which are valuable in telling us about changing weather. They can tell us when cold air is coming in, or a warm front is approaching. There are three main families of clouds: the cumulus, including the towering cumulonimbus thunder clouds; the stratus, layer clouds of middle altitudes, which cover the whole sky; and the cirrus, icy high clouds whose spreading across the heavens tell of a weather change.

Water vapour forms from evaporation by the Sun from the ocean's surface. It also collects through transpiration from plants, particularly from the equatorial rainforests. The vapour is mixed through the lower atmosphere (up to 12 km) by turbulence. There is very little water vapour or ice above this height.[5] The main role of water in the atmosphere is as a greenhouse gas, contributing up to 60% of the total greenhouse effect.

There are several techniques of obtaining water from the air, the most ancient of which are the dew ponds, which are not connected to groundwater or designed to catch rainfall, yet they always contain some water, condensed from the night air. They were made typically on the downs of southern England, insulated so that they remained cooler than the sun-warmed earth. They were often 20 metres *(70 ft)* wide, and 1.2 metres *(4 ft)* deep.

The optical effects of light on water vapour and ice particles can be extraordinary. The most striking of these is the rainbow, caused by refraction of light, with the same effect as a glass prism, breaking up the light into the colours of the spectrum during a shower with the sun behind you.

Other effects are rings around the Moon or Sun through high cloud, sometimes with a full coloured spectrum. From a mountaintop in winter you may see the 'glories' — a shadow of yourself projected by a low sun on to the cloud below, with coloured rings around your head. The better-known Brocken Spectre is similar, but with a longer distance to the shadow. These coronae depend on ice crystals rather than liquid water.

Cloud forms from surplus moisture in fully saturated air. The saturation level is called the dew point, which depends on temperature. Warm air can hold twice or three times the amount of moisture (and energy) than cool air, so you may get thunderstorms from warm saturated air and only drizzle from cool.

Condensation in the air rarely happens without some impurity present upon which the droplet can form. The air particles may be dust, smoke particles, salts or other 'wettable' nuclei. On land, condensation will form as dew on any object or plant, which means it is pure water. There is much folklore about the magical qualities of dew versus rain.

Cloud cover reflects the Sun's rays. The percentage of reflection is called the albedo effect. It can be as high as 80% from high cloud or from a surface of ice or fresh snow, and is an important secondary effect in global warming.

Clouds have always fascinated people. It is very therapeutic to lie on one's back in a summer meadow watching how the fluffy, constantly changing clouds form themselves into recognizable shapes of animals or faces. The ground, however, provides only a two-dimensional view. Better still to go up in a plane and wonder at the towering cumulus.

I found that gliding, which depends on finding updrafts of warm air, taught me so much about atmospheric air movements. It is fascinating to see how layers of clouds of varying height can move in different directions. Clearly there's a lot more going on in the atmosphere than meets the eye. It is a very complex system of air currents, with air masses of different temperatures and humidity that sometimes clash and produce dramatic weather.

Atmospheric circulation

The atmosphere acts rather like a gigantic heat engine constantly seeking to balance the temperature difference between the equator and the poles. Rising and descending air converts the heat energy into kinetic energy to provide the horizontal motion of the air streams within the troposphere (whose upper boundary is 9–18 km above the Earth's surface).

The wind belts and jet streams circling the planet are steered by three cells, the Hadley, Ferrel and Polar cells (see Figure 4), which are separated by boundaries of calmer air. The Hadley cell is an active, closed loop system extending from the Equator to latitudes 30° N and S, where the air descends, creating an area of high pressure. Some of this descending air moves along the surface, creating the trade winds. The Hadley cell moves further north in the northern summer and vice versa in the southern.

The Polar cell is also a fairly simple system, with warm moist air rising at about 60° latitude up to the top of the troposphere, moving towards the pole and sinking down to create an area of high pressure and the Polar easterly winds in the North and westerly near the southern pole.

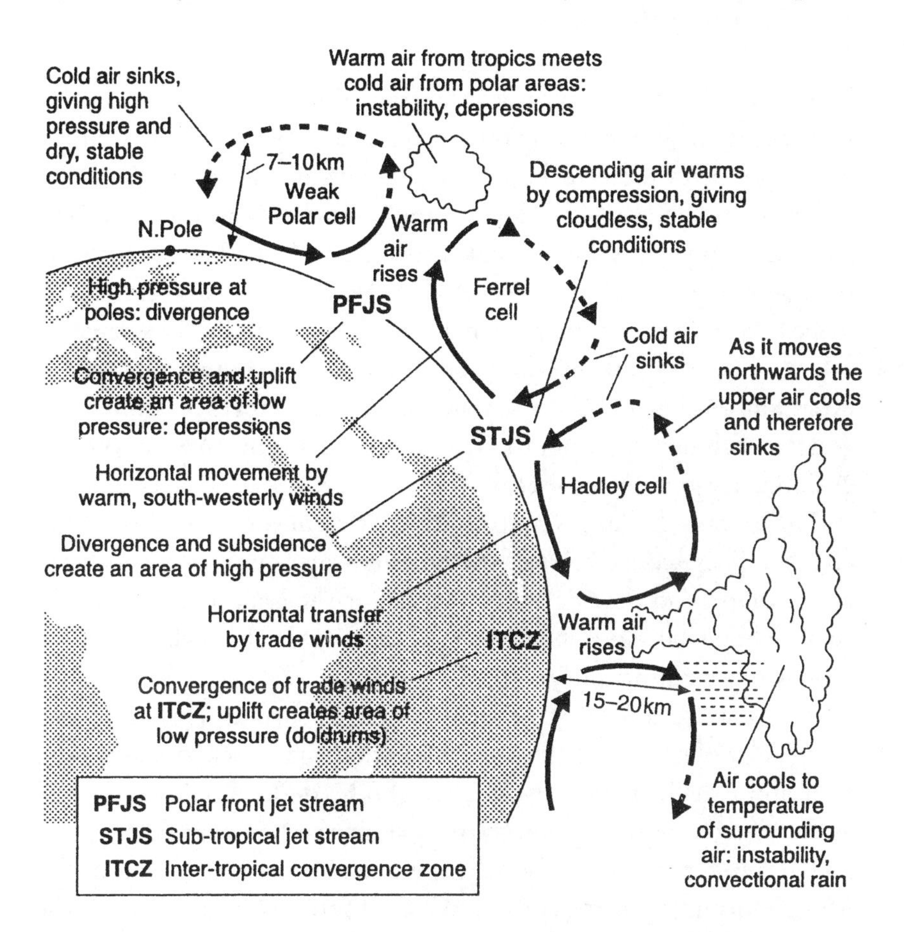

Figure 4. Atmospheric circulation in the northern hemisphere. The cells are formed by heated air rising, transferring heat from the Equator to temperate latitudes, with a second convergence and uplift at Mediterranean latitudes. Without this heat transfer (aided by oceanic circulation), the Tropics would be 14C° hotter, and the Polar regions 25C° colder. (Peter Bunyard)

Between these two circulatory systems is an area of more variable circulation called the Ferrel cell that acts like a ball bearing between the other two systems. While the upper winds and the jet stream will be prevailing westerlies, the lower air masses are influenced by high and low pressure areas that can cause large variations in wind patterns.

The varying path of the polar front jet stream makes a large difference to the weather. Its usual path with south-westerly winds brings most of Britain's more miserable wet weather. When it shifts northwards, Britain is bathed in hot, dry weather, when regions to the south are drenched in unseasonable rain.

Circulation of the oceans

The oceans were formed early in Earth's history, around $3\frac{1}{2}$ billion years ago, probably achieving their present salinity level about 1 billion years ago. They were to become the womb of life, from about 1.3 billion years ago until Caledonian times, some 400 million years ago, when the first plants and animals appeared on land. The oceans soon became the principal absorber of CO_2 from the atmosphere, which helped the planet cool and prepare for biodiversity of life. This ongoing role as a carbon sink has probably been its most important role in evolution. Marine life became abundant within a further 500 million years, absorbing CO_2 from the water. The oceans nurtured a complex evolutionary journey for sea life.

The oceans basically control the environment and the world's climate. They receive most of the Sun's energy as well as energy from the Cosmos, and provide water vapour for the atmosphere. The fact that we know far more about the land than we do about the ocean is a serious shortcoming today.

The basic oceanic circulation is driven by surface winds and also by the Coriolis effect of the Earth's rotation, anti-clockwise as viewed from above the North Pole (see Figure 5). This deflects fluids to the east as they flow from the equator to the poles in the Northern hemisphere, and to the west in the Southern. It is initiated in the area of the Gulf of Mexico forming a strong current, the Gulf Stream, which flows up the eastern side of North America to Newfoundland, and then crosses the Atlantic towards Western Europe. The warmth of this current raises the temperature of the British Isles by about 4C°.

As it turns west to the south of Greenland, it meets the cold, saltier waters coming down from the Arctic Ocean, which act like a pump pulling the stream into the deep, where it returns to the Southern Atlantic and then into the Indian and Pacific oceans as a slow, bottom current. In the Northern Pacific it wells up and returns as a surface flow, back through the Indian and South Atlantic oceans to complete what is called the Thermohaline Circulation System, with branches to the southern oceans. The complete circulation takes as much as one thousand years to complete. (See Plate 1.)

There is much concern with recent research which shows that the Gulf Stream's flow rate decreased by 30% in the thirteen years between 1992 and 2005, caused by fresh meltwater from Greenland's icecap, which is believed to interfere with the circulation by stopping the Gulf Stream's waters sinking.

It is thought that a slowing or shutting down of the Gulf Stream has happened a number of times in the historical past, bringing very cold conditions to Western Europe and to the east coast of North America. These climatic changes can happen suddenly and may last for a century or more.

Monsoons and tsunamis

A combination in the summer of strong evaporation from the ocean with land warming can draw in very humid air masses in the tropical and sub-tropical latitudes. The southern Asian summer monsoon is initiated by dry air rising from the warming Tibetan high plateau, which pulls the humid air in from the Indian Ocean. As we have seen earlier, the enormous Himalayan mountain range causes these moist air masses to rise and deposit vast amounts of rainfall, as the monsoon.

The oceans can produce devastating damage to coastal regions with a long-range wave pattern called a tsunami, caused usually by underwater seismic activity. These wave systems can travel at speeds of over 800 kph *(500 mph)* across thousands of miles of ocean, slowing as they approach land. Historically the damage caused when they hit a landmass was greatly lessened by the mangrove swamps that considerate Nature placed along the low-lying shore lines, acting like a shock absorber.

Technological Man does not understand their significance, and has steadily removed these safety barriers in order to establish profitable shrimp farms and rice paddies. The enormous loss of life in the December 2004 Indian Ocean tsunami and the 2008 Burmese typhoon would have been mostly avoided had these mangroves been left intact. Similar lessons were not learned, either, before hurricane Katrina struck New Orleans in 2005.

The terrestrial water cycle

The underground water cycle is a vital part of water's story, creating aquifers and huge storage systems that have remained, even under deserts, for many millions of years (until technological Man, without a thought for the future, is now draining them unsustainably). Wells and springs are part of this system. The relationship between underground and surface water cycles is significant. Working together, they form a balanced system of fresh water of high enough quality to allow the enhancement of life, providing the optimum conditions for biodiversity. In our ignorance and greed today, we have destroyed this balance, most cynically and critically in our destruction of the tropical rainforests.

This combined cycle of water, minerals and trace elements, kept active for millions of years by occasional orogeny, allowed nutrients to penetrate the banks of rivers, to create fertile flood plains and, with the cooperation of plants and bacteria, gradually to build up a soil profile, often many metres in thickness. From the onset of each episode of mountain building, it might take scores of millions of years to establish the soil fertility required for abundant growth and forests. The forest was Nature's brilliant innovation for the next surge of evolutionary expansion, at its most developed in the tropical rainforest. Viktor Schauberger demonstrates that the forest is also the cradle of water.

A part of the water cycle that is crucial in creating a climate that allows life to evolve fruitfully with increasing complexity, biodiversity and quality is the role played by vegetation by producing oxygen through photosynthesis and highly energized water through transpiration.

Trees do this most efficiently, and the tropical rain forests have huge impact on the world's climates (see Chapter 8). The greatest tragedy affecting the future of humanity is the collusion of the world's leaders with greedy commercial interests to destroy the rainforests. We hear much of the concern felt by many groups for the loss of species, especially of valuable plants that have not yet been properly studied or collected. This is essentially a secondary consideration, the primary being the climate, without which species biodiversity cannot exist. However, the damage caused to the world's forests, which will seriously exacerbate climate change and global warming, is barely acknowledged. We discuss the broader effects of climate change and water scarcity in more detail in Chapter 17. (See also Plate 2.)

The full hydrological cycle

In the same way that blood flows through the arteries and veins of the human body, so does water flow through the lithosphere of the Earth. The cyclical movement of water from subterranean regions to the atmosphere and back again is called the terrestrial water cycle.

The diagram below (Figure 5) shows the full hydrological cycle. Fresh water evaporates from the sea, rises, condenses and falls as rain. Some sinks into the earth and some drains away over the ground surface, depending on whether the ground is forested and what type of temperature gradient is active. In areas of natural forest where a positive temperature gradient normally prevails, about 85% of rainfall is retained, 15% by the vegetation and humus and about 70% sinking to the groundwater aquifer and underground stream recharge to pick up the negative energy charge of the Earth.

In a natural forest, the mature trees with deep roots bring up this negatively charged water, along with vital minerals and trace elements from the deeper soils. Trees act as biocondensers, harmonizing the positive energy from the Sun with the negative energy of the Earth (see Chapter 8). As a result, the evapo-transpiration from the leaves of the trees is a balanced, creative energy.

The forest, as a more dynamic living system, creates transpiration that carries the subtle energy (non-material) imprint of all the resonances of the complex biosystem, including the subterranean elements. Rainfall generated from the forest carries this beneficial influence. The ocean, although it is recharged by undersea volcanic eruptions and exposure to the atmosphere, mainly consumes all it produces, and therefore lacks these dynamic qualities. This is best explained in terms of homeopathic theory, in which the greater the dilution of a substance, the more powerful its energetic effect. This is an aspect of the ability of water to carry information that we shall be exploring later.

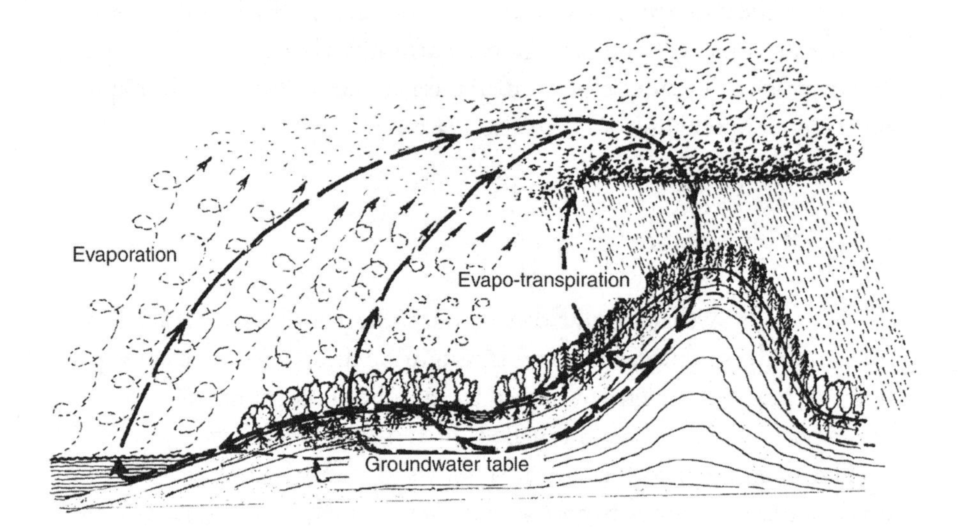

Figure 5. The full hydrological cycle. Water vapour accumulates in the atmosphere from evaporation from the ocean and from evapo-transpiration from plants. A positive temperature gradient allows absorption of rain into the ground, replenishing groundwater supplies. (Callum Coats)

The reduction in evapo-transpiration from the dynamic forests substantially affects the quality of the water vapour and its distribution in the atmosphere. The water vapour created by the natural forest has been balanced by fertile energies from the Earth that bring with it the power to stimulate and heal. Water vapour from the oceans has more of the raw untamed energy of the Sun, and global warming increases the evaporation from the oceans. Without the forest's water, there is a greater contrast between areas with abundant water vapour and those with almost none. This greatly disrupts weather patterns, with an increase in violent storms, hurricanes and serious flooding near coasts, while the areas away from coastal winds suffer droughts and cold night temperatures.

The half hydrological cycle

Man's clearing of trees and ground cover allows the land surface to be exposed. Without forest cover, the ground surface overheats, causing a negative temperature gradient in the soil. This means that the cooler rain cannot penetrate into the warmer ground, and fast surface runoff in areas of heavy rainfall causes catastrophic floods. The cause of the floods in recent years, in Central America, Columbia, Mozambique, Assam and Bangladesh was the deforestation on high ground.

This disruption of the natural water cycle, which Schauberger called the half hydrological cycle, is now prevalent almost worldwide; its shortcomings contributing significantly to our present climate change. Notice the difference between Figure 6 below and Figure 5 above. The drawing below shows that, in the absence of tree cover, the water table has sunk. Once the forest has been removed, the exposed ground heats up rapidly, all the more so if dry.

This type of evaporation, now lacking the evapo-transpiration from living things, has more destructive energies. If the rainfall is excessive, then flooding inevitably occurs. In many hot countries denuded of vegetation, dry valleys and creeks can be suddenly engulfed by a wall of water as terrifying flash-floods sweep away everything in their path.

In the absence of trees and ground cover to absorb it, the rainwater spreads widely over the surface of the ground, resulting in massive abnormal re-evaporation. The increase in water vapour in the atmosphere soon causes increased precipitation. What happens is that one flood causes another, while in inland areas, droughts become more frequent. The only answer to this vicious cycle is a massive international campaign to plant trees, particularly in the warmer latitudes.

The most serious result of the half cycle is that there is no replenishment of the groundwater. With the sinking of the groundwater level, the supply of nutrients to the vegetation is cut off. The water that is evaporated into the atmosphere is virtually lifeless, lacking in the energy and the qualities that groundwater acquires. Viktor Schauberger called this a 'biological short-circuit'. The essential soil moisture, trace elements and other nutrients that the tree roots normally raise to the benefit of other plants sink below reach as the groundwater sinks. This is the cause of desertification, now becoming prevalent in many tropical areas. The groundwater disappears, probably for ever, into the womb of the Earth where it came from.

The limited circulation of the half water cycle also increases the intensity of thunderstorms. These can raise the water vapour to levels far higher than normal. At altitudes of 40–80 km it is exposed to much stronger ultra-violet and high-energy gamma radiation, which break up the water-molecule, separating the hydrogen and oxygen atoms. The hydrogen then rises because of its lower specific weight, and the oxygen sinks. That water becomes permanently lost. The effect of global warming is complex. The atmosphere first warms up due to the greater amount of water vapour, some of this increase of heat being offset by the loss of water atoms at high altitudes.

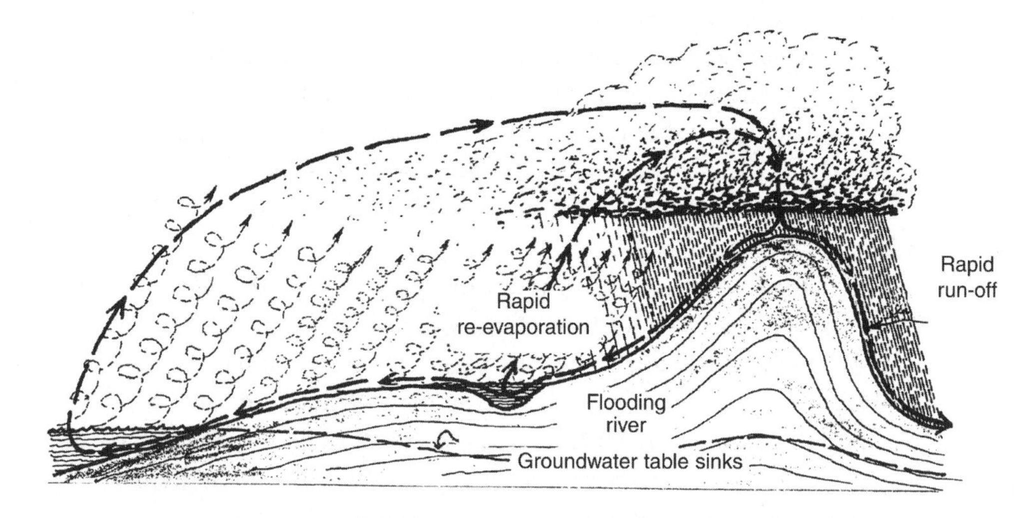

Figure 6. The half hydrological cycle. Because the land surface is warmer than the rain (negative temperature gradient), it sheds the rain, causing rapid run-off and flooding. The groundwater table sinks. (Callum Coats)

6. Springs and Rivers

When you drink the water, remember the spring.
Chinese proverb

Before the installation of public water supplies, springs were the most valued or sometimes the only sources of drinking water, and they still are in most parts of the world. Settlements would establish around a spring that delivered high quality water. Possibly because of the connection between living water and good health, some established a reputation for curative powers. Viktor Schauberger was convinced that the high quality water produced by his springwater machine had healing qualities.

Spring water and mineral water are often said to be 'pure'; but what is meant is 'good quality'? The term 'pure' should be reserved for H_2O, or virgin water, found in Nature only in evaporated form.

The veneration of springs

Springs have long been associated with folk medicine, ritual and religion, frequently being reported as places of power in the landscape (Plate 15). Often, springs thus endowed are called 'holy wells', which is confusing, because the word derives from the Anglo-Saxon for spring — *wella* (hence the expression 'to well up'), not for its modern use as a shaft excavated to reach the underground water table. The tradition of venerated springs is found in all cultures and major religions, going back to prehistoric times. The most common association is the bestowal of supernatural qualities, but more specifically as the abode of spirits or deities, or being linked with holy figures or saints. In Britain, in most cases, the saints named had no connection with the site, but their qualities may be associated with those previously ascribed by the pagan culture.

The waters of most sacred springs are credited with healing powers and with cures assisted through bathing or drinking. In British lore the most common affliction whose healing is claimed is infertility, followed by eye complaints. However, some springs are regarded as being so powerful, as at Lourdes in France, or England's only thermal spring at Bath, that they are reputed to heal many diseases. Offerings (usually coins) were made to the pools served by the springs, either as part of the locally established ritual, or as a 'trade' for a wish to be granted. Many 'wells' were 'dressed', or decorated with flowers, paintings, statues or strips of coloured cloth, a tradition found all over Europe and Asia, in Africa and Central America.

Natural springs would be valued also because the quality and reliability of the water flow in times of drought might make the difference between life and death. It is not hard to see why people invested these sites with magical powers, or saw them as inhabited by a living spirit who was the guardian of the waters. It is likely that many of our forebears would empathize with Viktor Schauberger's vision of water as 'the blood of the Earth' when they saw the perfect, cold, nourishing liquid issuing mysteriously from the womb of the Earth.

When the rationalism of the Enlightenment replaced the superstitions of an earlier age, some explanation had to be found for the curative powers of certain famous springs. This led, in the 18th century, to the birth of the spa culture, and doctors would examine any deposits left behind when they had boiled away the water, in order to identify this and that mineral as the true elixir that would give legitimacy to their spa water. During the Protestant Reformation in England, and then with the decline of rural populations, many sacred springs fell into disuse, being rediscovered in the nineteenth century by Irish immigrants whose Celtic-based Catholicism still had strong pagan roots.

Today, with the revival of ancient rural traditions, many sacred springs are being restored in Britain and in Continental Europe.

Seepage springs

What is generally called a spring is actually not a true spring, but a seepage spring, which is the overflowing of surplus water from soil and rock strata that have a limited depth. Rainwater that is warmer than the ground (a positive temperature gradient), soaks in and sinks until it reaches an impervious layer like clay which channels it out as a stream to the surface again, lower down. It acts by gravity. The temperature of the water will be that of the strata from which it emerges, probably between +6°C *(43°F)* and +9°C *(48°F)*. This water will contain some trace elements, minerals and dissolved salts but, generally speaking, not in such a broad spectrum as in true springs (see below). The seepage spring responds quickly to variations in precipitation, frequently drying up in a hot summer and flowing strongly after heavy rain.

True springs

A true spring originates from much deeper strata (see Figure 8). Water collects in ancient aquifers and retaining basins over many years, and the water emerging to the surface might be hundreds of years old; or even thousands in the case of famous therapeutic hot springs. Because of their age, these spa waters are extraordinarily rich in well-balanced minerals.

The rich waters of the Hunza Valley in Pakistan, or the Caucasus mountains, which are credited for the longevity of the local people, also originate in true springs. The difference here is that, emerging in the high mountains, these waters are then augmented by rich glacial waters, and by minerals from the action of the aggressive mountain streams eroding the surface rocks.

The rainwater penetrates the ground surface under the influence of a positive temperature gradient in a way similar to that of a seepage spring. But it is drawn down much more deeply, helped by the increasing pressure, so that it condenses and cools to around +4°C *(39°F)*. Being immature water, it will absorb what it can, so it removes salts from the upper layers of the ground, depositing them later as the water condenses and cools with depth. This makes the upper layers more fertile, and the salts are now available to deep-rooted trees that have the ability to metabolize them, converting them to nutrients for more shallow-rooted plants.

The downward-percolating rainwater increases the pressure on the groundwater body, pressing the lowest layer into rocks that are affected by geothermal heat. These are caused to expand, compressing the layers above. But the +4°C *(39°F)* stratum water is already at its densest and virtually incompressible at this temperature, so all it can do is to push out laterally, providing the springs with their flow. This action explains how springs can emerge from high mountain peaks at such cold temperatures, where there would be insufficient local collection for gravity.

Rain absorbs oxygen in its fall through the atmosphere. After it enters the ground and percolates through the soil, plant roots and organisms reduce its oxygen content. So when it eventually emerges as a true spring, the water is often oxygen-deficient, though rich in carbonic acid. It is unwise to drink this water directly from the spring, for being hungry for oxygen, the water can steal it from susceptible organs, like the stomach, causing great discomfort. If breathed directly,

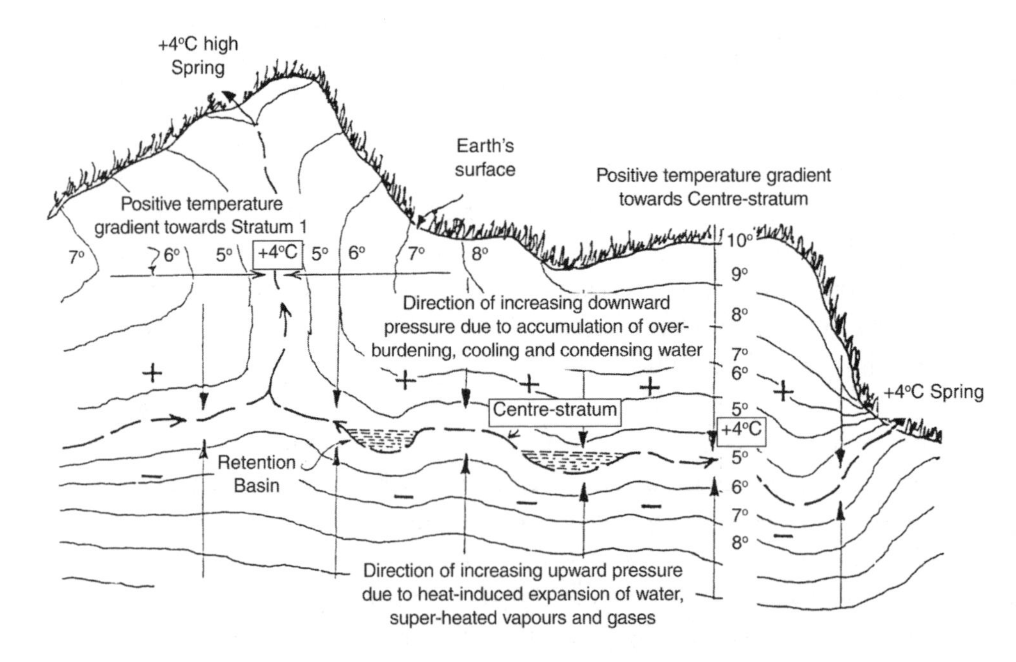

Figure 7. True Springs and high altitude springs. These depend on a +4°C (39°F) denser water level called the centre stratum. This gets squeezed between the weight of water in the rocks above and the strata below. At +4°C it will compress no more and moves vertically or horizontally, emerging as a spring. This is why they are normally very cold and may appear on mountain tops. (Callum Coats)

the carbonic acid can damage the lungs. Known to mountain folk as 'damp-worm', and by miners as 'choke–damp' respectively, both can be fatal. However, within ten metres of the source, the water has usually, through its active movement, absorbed sufficient oxygen to be quite safe to drink.

Rivers and how they flow

If we understood the importance of water both for the environment and for life, we would nurture and protect our rivers, which are the great arteries of the Earth. Healthy streams and rivers are water at its most active, powerful and playful. In our ignorance of how water needs to move, we restrict rivers with embankments and other unnatural constructions. We treat rivers as sewers for waste, and we extract the energy and spirit from their form in dynamos.

For scores of thousands of years, since people started to settle on the land, our forebears were aware that their prosperity depended on the river. Soils are quickly depleted of their nutrients by agriculture, particularly if intensive. Remineralization by regular flooding of the river was vital to obtaining good crops. This allowed the great civilizations to grow and flourish, in Mesopotamia, the valleys of the Nile, the Yellow River and the Indus, to name a few.

Today's technocrats have a need to control this apparently chaotic behaviour of the natural river, by steering the flow, sometimes behind high banks, and disregarding the eco-system, to the great loss of fertility of the surrounding fields. Modern NPK artificial fertilizers (nitrogen, phosphorus and potassium) cannot take the place of Nature's remineralization; in fact they often cause great problems through creating imbalances and pollution.

Stages of a river

A river has three stages of life. Its youthful stage energizes the water as the steep landscape puts it through vigorous tumbling, spinning and intense vortical movements. The immature cold water is hungry, taking up minerals as it scours the rock, cutting gullies with the suspended gravel and steepening the sides of the valley, more especially when it is in spate. It is oxygenated in rapids and waterfalls. It is put through exercises that it will use well when it matures.

When the stream leaves the steep country, the flow slows, and some of the heavier rock matter it carried in suspension is deposited, to be picked up again when the flow accelerates. The water is now mature, having absorbed minerals and generative energies, and if it is prevented from excessive warming by trees on its banks, the stream water is absorbed by the banks, recharging the groundwater of the surrounding countryside (see Figure 9). The richness of movement of the young stream is carried into the body of the meandering river. The water is creating its own form, which in turn regulates its flow.

On entering the plains, the river, in its natural way, would meander across the flat country, and when a bend twists back on itself, a shortcut will be created at flood time, leaving behind an ox-bow crescent lake. It is in the plains country mostly that people try to manipulate the river, heavy with silt, by straight embankments to stop the river spreading

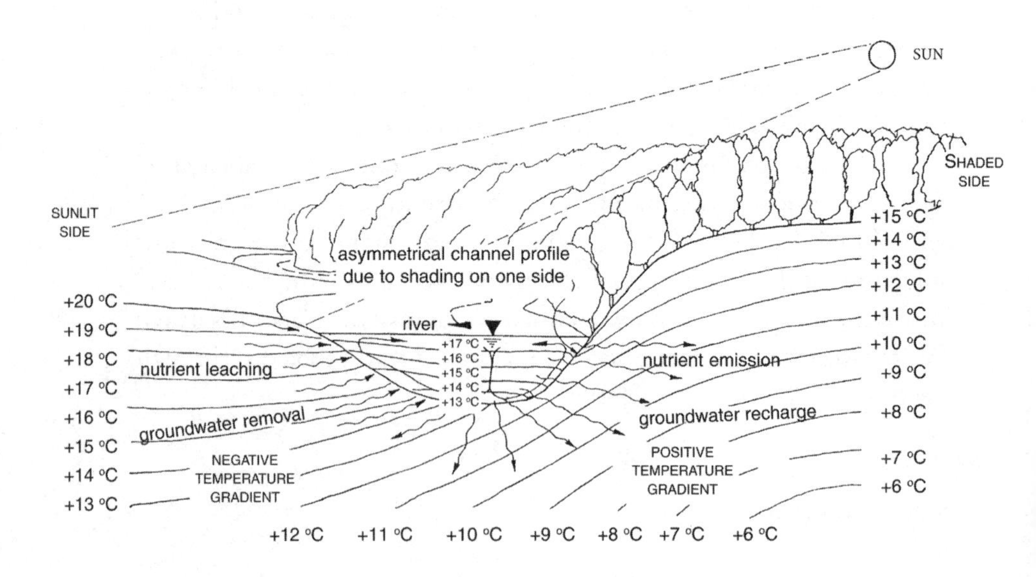

Figure 8. Asymmetrical river development. The bank exposed to the Sun is warmed so that its groundwater leaches minerals into the stream. The shady side allows the cooler ground to absorb the mineral-rich water. (Callum Coats)

where it wants to. If these natural floods were permitted, they might not be particularly destructive, and they would remineralize the soil, which becomes much more productive.

However, technical man believes he can control Nature. The old river is now typically forced to perch sometimes 50 feet above the surrounding countryside. If the river should burst its artificial banks at this stage, the flooding is often catastrophic. Lacking its normal twisting movement and the positive temperature gradient that keep the silt in suspension, this is deposited, blocking the channel. Its natural path thus obstructed, the flow becomes angry and unpredictable. There are now very few major rivers that are allowed to flow naturally.

How the river protects itself

Schauberger saw water as being conceived in the cool, dark cradle of the virgin forest. As it slowly rises from the depths, water matures by absorbing minerals and trace elements on its upward path. Only when it is ripe will it emerge as a spring. A true spring (compared to a seepage spring) has a water temperature of about +4°C *(39°F)*. In the cool, scattered light of the forest, water begins its long journey down the valley as a lively, sparkling and gurgling stream.

Water, when it is alive, creates this spiralling, convoluting motion to retain its coolness and maintain its vital inner energies and health. It is thus able to convey its vibrancy to the surrounding environment. Have you noticed how refreshing and enlivening it is to sit by a healthy bubbling stream?

Naturally flowing water seeks to protect itself from the damaging direct light of the Sun. The reason that you find trees and shrubs growing on the banks of streams is not from people planting them, but because the energies from the flowing stream facilitated their growth there, to shade the water. When a stream is able to maintain its energies, it will rarely overflow its banks. In its natural motion, the faster it flows, the greater its carrying capacity and scouring ability and the more it deepens its bed.

Schauberger discovered the reason for this — that in-winding, longitudinal spiral vortices form down the central axis of the current, moving alternately clockwise and anti-clockwise (see Figure 12). The nature of inwardly spiralling vortical movement is to cool. So these complex water movements constantly cool and re-cool the water, maintaining it at a healthy temperature, leading to a faster, more laminar, spiral flow, ejecting or neutralizing undesirable substances.

As the stream gets bigger, it is less able to protect itself from light and heat, and it begins to lose its vitality and health, and with this its ability to energize the environment through which it passes.

Ultimately becoming a broad river, the water flows more sluggishly and becomes more opaque with the increasing silt content makes. This, however, protects the lower strata from the heat of the Sun. They remain cooler, retaining the spiral, vortical motion that is able to shift sediment of larger grain-size (pebbles, gravel, and so on) from the centre of the watercourse, and keep down the risk of flooding. This motion also discourages the generation of harmful bacteria and the water remains disease-free.

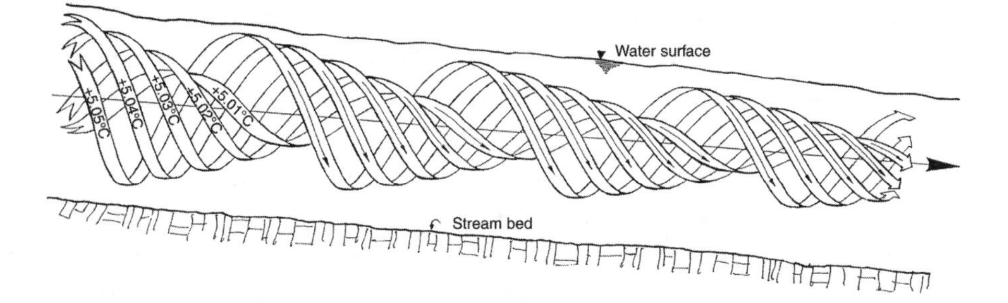

Figure 9. A longitudinal vortex showing the laminar flow of a river around the central axis where the water temperature is always coldest. The central core displays the least turbulence and accelerates ahead, drawing the rest of the water body in its wake. (Callum Coats)

Temperature gradients and nutrient supply

As we have seen, unless vegetation keeps the ground surface cooler than the falling rain, the water will not easily penetrate the soil. The direction of the temperature gradient indicates the direction of movement. Energy or nutrient transfer is always from heat to cold. So a positive temperature gradient is also essential for nutrients to be able to rise up to the roots of the plants (see Figure 10a).

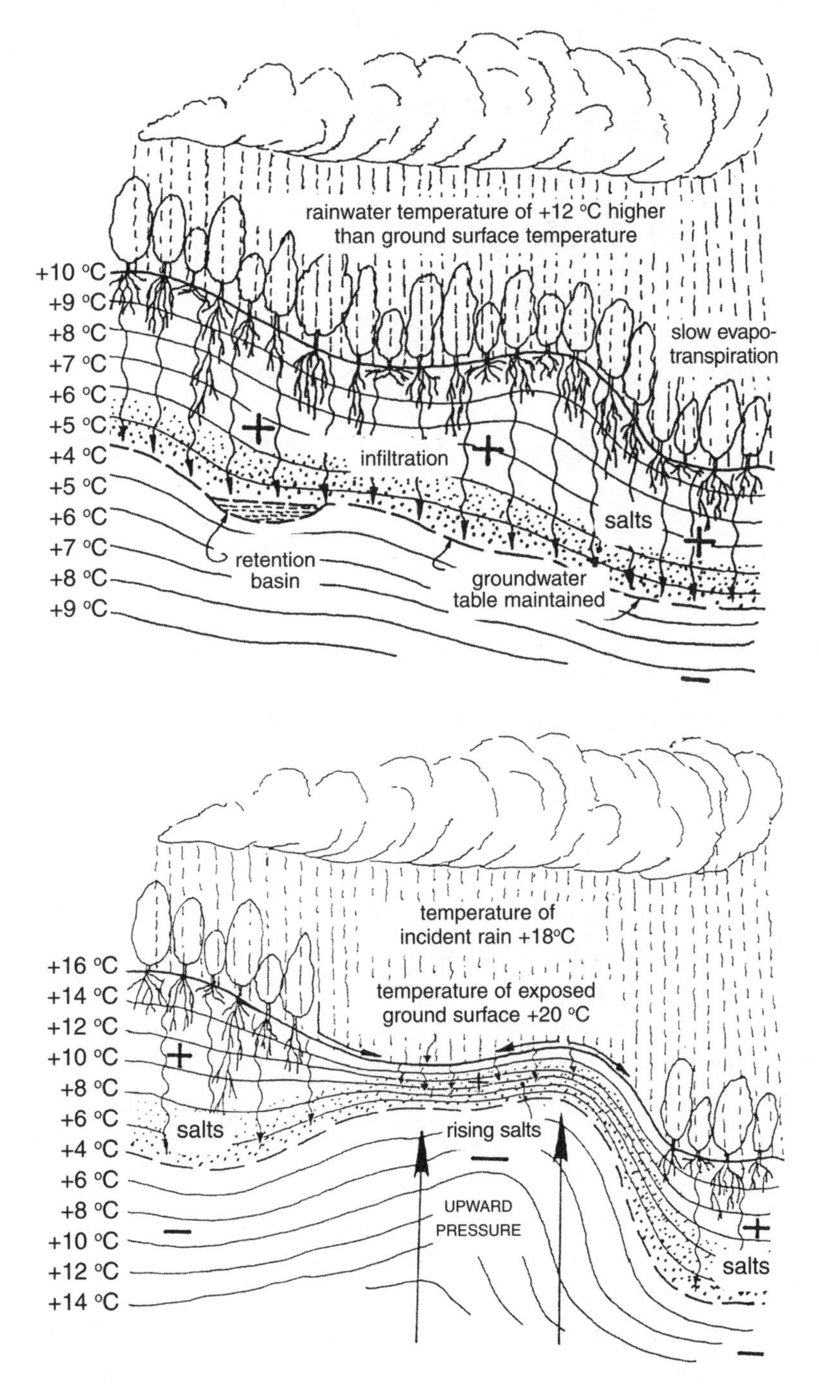

Figure 10. Temperature gradients. Upper: The warmer rain is absorbed into the cooler ground (positive temperature gradient) as it is protected by trees. Lower: The unprotected land surface sheds the cooler rain, causing the groundwater to rise and with it, unwanted salts. (Callum Coats)

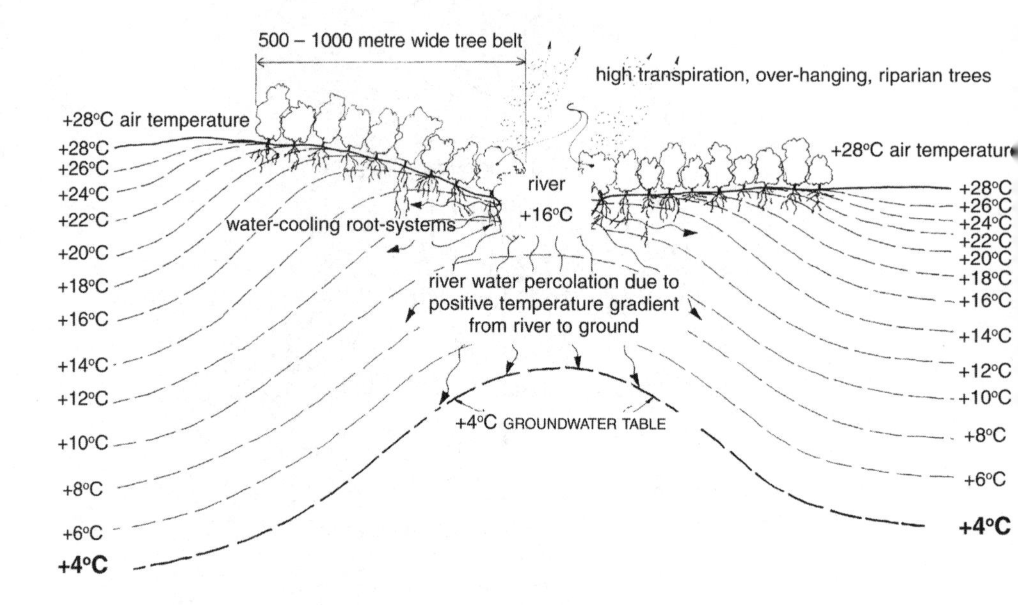

Figure 11. Groundwater recharge. Trees cool the ground like a refrigerator, allowing a positive gradient to draw water from the river to recharge the water table. (Callum Coats)

If the surface is well forested, the rainwater is warmer than the soil, and penetrates to the lower strata, replenishing the groundwater body and the aquifers. The salts remain at a level where they cannot pollute the upper strata when they would harm those plants that are salt-sensitive. The groundwater hugs the configuration of the ground surface. Figure 11 shows how the salts in the ground rise near the surface, particularly on a hilltop, when part of the forest is cut down, leaving the ground exposed to sunlight.

Schauberger demonstrated that when light and air are absent well below the surface of the ground, the minerals and salts are precipitated near the temperature horizon of +4°C *(39°F).* Warm ground will encourage evaporation of the moisture near the surface, so that the minerals and salts are deposited near the surface, lowering the fertility of the soil. If all the trees are removed, there will be no penetration of rainwater; the water table initially rises, due to the now uncompensated upward pressure from below described in the chapter above, bringing up all the salts, but will eventually sink or disappear altogether without the replenishment of rainwater. Fertility can be restored in time only through reforestation, bringing about the reestablishment of a positive temperature gradient.

Replanting must be done initially with salt-tolerating trees and other primitive plants, as only these would survive under such conditions. Later, due to the cooling of the ground by the shading of the pioneer trees, the rainwater can penetrate the ground, taking the salts with it. Over time, as the soil climate improves, the pioneer trees die off, because the improved soil conditions don't suit them. Other species of tree can replace them and the dynamic balance of Nature is restored.

Irrigation in hot climates aggravates the problem because, as the ground temperatures cool during the night, the irrigating water can penetrate the upper salt-containing strata. With the increase in temperature during the day, the infiltrated irrigation water with its acquired salts are drawn up, and upon exposure to light and heat are deposited on the soil surface. The seriousness of the problem will vary with latitude, height and season.

The formation of vortices and bends

We have seen that energy is always connected with movement. The natural movement of water is sinuous, convoluting and vortical. Without such movement there is no polarity. Vortices, however, cannot form without the existence of polarities. Through the action of vortices come rhythms, the pulsations that act as a gateway — a breathing process that the river performs for the environment.

Schauberger invented a novel way of deflecting the flow to the centre of the stream in order to form cooling longitudinal vortices.

Hydro-electric power

Until the development of coal-fired steam engines, water wheels were an important source of power. With the arrival of electricity came an interest in water as a large scale energy source, through the building of dams to provide a higher level of kinetic energy. There are now tens of thousands of hydro-electric dams which collectively generate about twenty percent of the world's electricity. Dams prevent downstream flooding and provide water for irrigation. However, they are controversial because the associated upstream flooding destroys

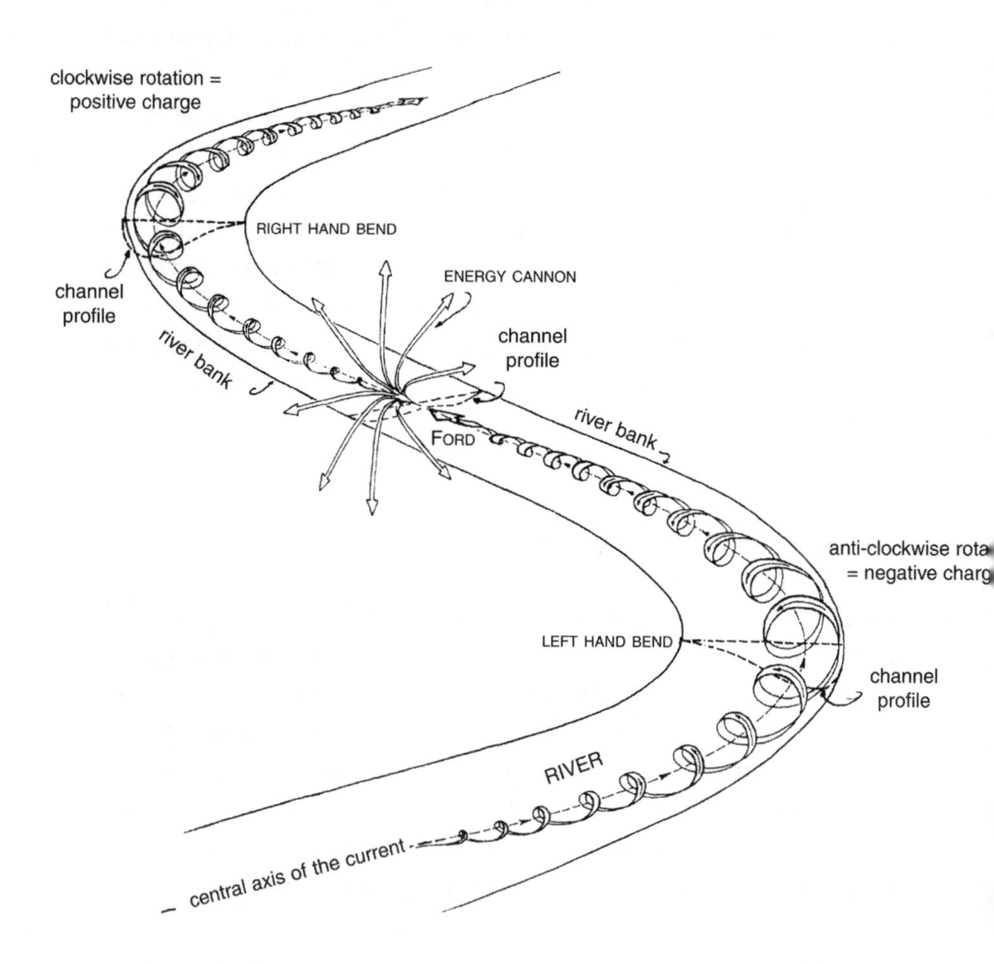

Figure 12. The energy cannon. Approaching a bend, the water forms a cooling anticlockwise longitudinal vortex (see Figure 9 above), grinding sediments and releasing nutrients into the river. After the bend, the vortex slows down, and the water warms in the shallower riverbed, releasing nutrients. Just before a new clockwise vortex forms anticipating the next bend, it releases energy into the environment. Schauberger called this the 'energy cannon'. (Callum Coats)

valuable agricultural land, ecosystems and communities that have been established for centuries. If there is any deforestation or agriculture upstream, the artificial lake can silt up quickly, making the whole project virtually useless. Hydro-electricity is often seen as a desirable source of power as it does not produce CO_2. However, the production of every tonne of cement creates a tonne of CO_2 and the steel materials and transport required for construction create a whole lot more.

Large dams are built largely for the political and economic interest of central government and multinational companies. They can have disastrous effects on rural populations. In India, notably, large dams have displaced between 16 and 38 million people.[1] The Three Gorges hydro-electric project on the Yellow River in China is particularly controversial. It has taken out some of the most fertile land in central China. Here, 1.24 million people have had to be resettled, usually on poor quality land, the river's fauna has been irretrievably damaged, archeological treasures submerged, and already the enormous lake is silting up. All this is to satisfy the rapidly expanding and energy-hungry Chinese economy. It could become a catastrophe, particularly if the enormous dam was damaged by an earthquake, a not entirely unlikely event in this part of China. The Yellow River's energy dominates central China. Its interruption will affect the whole country.

At present (2010), details have not yet been published on the World Bank financed River Congo dam which is three times larger than the Three Gorges project. The health of the tropical rainforest is closely dependent on its main river system. If this is emasculated by a gigantic dam, the whole of central Africa will suffer irremediably.

The prevailing methods of hydro-electric power generation destroy the water which is thrust down cylindrical pipes under enormous pressure.[2] Upon leaving these, it is then hurled against steel turbine blades where it is smashed to smithereens. The physical structure of the water is literally demolished and all the dissolved oxygen, and even some of the oxygen in the water molecule itself, is centrifuged out of the water.

This fragmented and largely oxygen-deficient water, a virtual skeleton of healthy water when forcibly expelled into the river, has disastrous consequences for the fish and other aquatic life. Inevitably certain species of fish disappear once these power stations are commissioned, and other forms of life survive with difficulty.

Smaller scale and micro-hydro are more environmentally friendly. Some even have installed fish ladders so that the energy created by the stream which nourishes the valley is not completely destroyed. Viktor Schauberger and others produced plans for channelling water spirally, so that its energy is not damaged by the generators.

The water that springs from the mountains is the blood that keeps those same mountains living and is the vein that is formed within and across them.

LEONARDO DA VINCI

7. Water and the Human Body

Water is the bridge between the silent, unseen ocean of electric and magnetic vibrations in which we are now embedded, and the subtle, complex biochemical rhythms that form the basis of all organic life.
Alan Hall, *Water, Electricity and Health*

Albert Szent-Gyorgyi, who discovered vitamin C, called water the mother and matrix of all life. He said that water is so much part of life that we tend to ignore it and look elsewhere for the magic bullet, the secret herb or nutrition that will increase health and energy and extend life-span.

Biological water

You know that the human body is composed of 50–80% water. The actual amount varies with age and the amount of fat. For example, for a foetus in the womb water is about 90% of total weight. This drops from about 74% as an infant to 60% as a child. Muscle stores fluid (75% of muscle tissue), so adult men tend to have greater water content than women, but the amount of water gradually drops with age, so that in the elderly it may drop to below 50%.

Water's journeys through the body amount to an incredible complexity of inter-dependent relationships; studying them, the impression steadily grows that water truly behaves like an organism. A list of the bodily fluids numbers some thirty, each with its specific purpose.[1] Most of us do not drink a sufficient amount of good quality water to maintain the metabolic action and the flushing out of wastes and toxins at their optimum level. What is the function in the body of all this water?

—A significant proportion of it is 'bulk water', found in the principal organs which have specific metabolic functions to do with energy exchange.
—Water drives the digestive system; it is the basis for the digestive juices and acids that break down our food.
—It drives the respiratory system: the lungs and sinus systems depend on large quantities of water.
—It is essential to our circulation system: our blood is largely water: 92% of blood plasma; 60% of blood cells. The kidneys, another water organ, regulate the composition of blood and produce urine.
—For the liver's function of producing and converting nutrients from the digestive system and particularly for detoxification, water is essential.
—Because the brain has a high water content (about 85%) adequate hydration is required for efficient brain activity.

Cellular water

The greatest amount of biological water is extra-cellular water. It is not pure water but has a saline content, which allows it to carry an electric charge, for it has the essential role of facilitating instant communication across the whole organism.

One of the most important functions of intra-cellular water is to facilitate cellular functions in the body. Because of its unusual hydrogen bonding, it has the unique ability through hydration to activate proteins, and through its ionizing abilities to facilitate proton exchange and cell formation. Its particular ability to act as a solvent is essential in the action of salts and ionic compounds. Extra-cellular water supplies nutrients and removes toxins.

Human blood

The blood of the human body is 90% water. Its composition is comparable to sea water, except that the ocean's principal salt is potassium, the blood's is sodium. It was believed by scientific intuitives like Goethe, Steiner and Schauberger that blood behaves like an organ.

It is common for those who use the ancient practice of observing the breath when they meditate to experience the strange sensation of 'being breathed'; this experience seems to be part of a 'greater breathing'. Viktor Schauberger would often insist in a similar vein, that a bird 'is flown' and a fish 'is swum'. On many occasions he said that the heart is not a pump, that it 'is pumped'. He saw the heart, rather than as a pump itself, as a regulator of blood flow. The spurts of blood that the heart produces during contraction are more like the automatic reaction to having been full, like the outbreath from the lungs.

The humours — variations of biological water

The notion that health was controlled by the four humours dominated European and Arab medicine for 1800 years, from the time of Hippocrates (fifth century BC) until the mid nineteenth century when medicine moved away from the study of the body as a whole. They were the bodily fluids that were associated with specific organs: phlegm with brain and lungs; yellow bile with the gall bladder; black bile with digestion and spleen; blood with the heart.

They have correspondences: winter (wet and cold) is associated with phlegm and the element of water; spring with blood (wet and hot) and air; summer (dry and hot) with yellow bile and with fire; autumn (dry and cold) with black bile and earth. They were also associated with different temperaments: phlegmatic (conservative, compassionate); sanguine (optimistic, impulsive); choleric (ambitious, passionate), melancholic (creative, depressive).

It was a holistic theory. Illness occurred when the humours were out of balance. They were brought back into balance and health by eating the right foods or by removing the excess of one humour. Blood letting was common, as blood was thought to contain the other three humours. Vomiting was induced for someone suffering from a choleric complaint, diarrhea for melancholy.

The humours were a complex and sophisticated system, especially after being revived by Galen in the second century AD. It gave correspondences with temperaments, with astrology, with climates, latitude and environments. There were variations, such as five or up to twelve humours, but the four remained the most used. This system has continued, up to modern times, to influence the study of human behavior based on physiological origins. A number of modern researchers have based their analysis of temperaments and psychological types on the humours.

The body's skin

The skin of any organism performs a number of important functions. As the outside layer, it defines the integrity and coherence of the organism and protects its vulnerability to physical assault and infection. It is the vital heat-balancing organ for most animals, excess heat being taken off by evaporation through sweat glands. Sweating can also be a sign of stress. Lie detector tests measure the sudden increase of the water content of the skin (see Chapter 14).

The skin is porous. It can absorb fluids, not only therapeutically, as with aromatherapy, but also as creams to make it more supple. While protecting against invasion by harmful substances, it is sensitive to them, and can provide a useful advance warning system.

It is full of tiny sensors; in some animals these represent their main 'antennae' for picking up information from the surrounding environment. In humans, this function has become less important — though blind people depend on it — but we can still receive 'impressions' with the whole of our body (for instance, in dowsing). Our body is also the antenna for receiving deeper energies, cosmic, volcanic or earth energies. The skin's health is important for effectively performing these functions.

Water in the body

Water is an excellent solvent and suspending medium. In the blood it forms a solution with a small part of the oxygen you inhale, to carry the oxygen to body cells. It also dissolves some of the CO_2 that is carried to the lungs to be exhaled. Water moistens the air sacs in the lungs to allow inhaled oxygen to be dissolved and then distributed to different parts of the body.

In recent years research has demonstrated the importance of adequate hydration of the body. Modern diets, which include processed and adulterated foods, a shortage of fresh fruits and vegetables, usually too much sugar (or artificial sweeteners) and salt, commercial fizzy drinks with questionable ingredients (E numbers), too much tea and coffee (which are diuretics, or biological water reducers) — all these can lead to imbalances in the body, one of the results of which is dehydration.

Many of the body's primary organs depend on plentiful, good quality water for their efficient functioning, especially the brain, lungs, kidneys and liver. There are two principal energy cycles in the body — the kidney and the liver cycles. If the kidneys suffer dehydration, for example as a result of chronic diarrhoea, the lungs are affected, and breathlessness can become a problem, which can then affect the heart. Liver function is later affected, and its function of feeding energy into the blood is further compromised by dehydration. Unless prompt action is taken, dehydration in these circumstances can result in kidney and liver damage. We can survive weeks without food, but will soon die without water. (Shipwrecked at sea, survivors can keep going a bit longer by recycling their urine.)

Karol Sikora writes: 'The commonest cause of death is not cancer, stroke, heart disease or earthquakes, but dehydration caused by diarrhoea disease in 15 million children a year. Investing £10 a life could solve the problem.'[2]

Avoiding dehydration

It is clear that water plays a major part in balancing the health of the body, and both our kidneys and bladder are essential water organs. When you think about it, with any stress, water is ejected. When we have a cold, we sniffle; with a fever or excessive heat we sweat; when distressed or our eyes are irritated, we weep. The lungs must be lubricated with water to function. Next time you have a head cold, observe how much imbalance there is in your head fluids. Or, after a bad bladder night, see how much rehydration you need!

We all know how dependent our mouths are on water. It is the same with all animals. If the mouth is not constantly well lubricated, the whole body is eventually affected. Anxiety or tension can dry the mouth; as also can a lot of talking!

There has been a big change in public awareness of the need for water top-ups. Nowadays, it is normal practice to have water jugs and glasses at conferences. It is interesting that this perception is changing social conventions. In Britain, while it is still the cup of tea or the glass of beer for refreshment, more and more people take a small bottle of water with them everywhere. In France bottled water is replacing wine, while in the USA, people are even drinking water in place of coke. The greatly increased level of toxins in our environment has increased the body's need for water to help flush them out.

We are urged to drink lots of water for our health, up to two litres a day. While it is true that we lose on average about 1 $^1/_2$ litres a day through sweating, breathing and urinating, we also absorb water as part of our food intake, particularly from fruits and vegetables. Carbohydrates are broken down into glucose and glycogen. The glucose is available for immediate energy needs, while the glycogen, which holds a lot of water, is stored in the liver and muscles, and is released when the sugar is needed for sudden exercise.

This water needs to be converted into cellular water, which is easy enough when you are young and healthy, but it gets harder when you age or are unwell, when you need particularly to drink more and better quality water. Dehydration is a common problem with the elderly, as one of the problems older people often face is the shrinkage of their kidneys. The charity, Action on Elder Abuse, arguing for stronger guidance on malnutrition, finds that the elderly are more vulnerable to infections, dizziness, confusion and falls as a result of dehydration. In care homes in Suffolk, England where residents are encouraged to drink 8–10 glasses of water a day, the number of falls has dropped sharply.[3]

Patrick Flanagan's water research

Patrick Flanagan's speciality was the liquid crystal structure of water. At the age of seventeen he had been named in *Life* magazine as one of America's top ten scientists. A protégé of Dr Henri Coanda, the father of fluid dynamics, he researched for clues to the longevity found in mountain communities like the Hunza in Pakistan, in Georgia in the Caucasus and Villacabamba in Ecuador. People who live in these regions and drink the water are known to live long, healthy, active lives; many exceed one hundred years of age.

For ages, man has made pilgrimages to these remote regions in order to drink their 'healing waters', the chief constituent of which seems to be a rare form of colloidal silica. Flanagan discovered that this water came from glacial melt and had very distinct physical properties of viscosity, heat and energy potential. Normal water depends for its cohesion on ionization. Hunza water does not; it is similar to inter-cellular water whose liquid crystal structure is loose enough to transport easily both toxins and nutrients, saving the system from clogging.

Flanagan researched for decades how to replicate this water and eventually, in 1982, came up with 'microcluster silica' whose electromagnetic properties encourage the water molecules to separate, making the water 'wetter'; or more easily absorbed by the body. (A similar effect can be obtained with a magnetic pad placed under a water container.) The surface tension is lower, facilitating hydration, nutrient uptake and toxin removal. It also is claimed to have the quality of pH adjustment in whichever direction you need. He now markets this special water as 'Crystal Energy'.

The seeds for this crystalline structure were colloids, particles in suspension or solution, too small to be seen with an ordinary microscope, charged with energy to attract disorganized water molecules. The water cannot retain its colloidal structure without an electric charge, which is easily dissipated.

Flanagan was particularly interested in biological water, and found that it contained a much higher percentage of structured water because the colloids had their electrical charge protected by a coating of collagen, albumin or gelatine.

From Theodor Schwenk's research (see Chapter 13 below), Flanagan knew that all flowing water, though appearing to be a uniform mass, is made up of layers. Turbulence in water produces a more complex laminar structure, generating an electrical charge difference between the layers. He wanted to see if he could reproduce the effect of the fast-flowing glacial torrents by constructing an artificial vortex, realizing that it was the stream's vortices that charged the particles, allowing them to become colloids.

To view the intricate structure of the vortex, it helps to add a little glycerine, or a few drops of food colouring. You will then see that it seems to have a life of its own — the diameter of the structure shrinking as the point plunges to the bottom of the vessel — expanding as the point rises up and the structure diminishes. There is a rhythmic pulsation, like breathing, that is the mark of a living vortex, whose formative inner layers spin more rapidly than the outer layers which develop corkscrew forms like the spirals inside conch shells.

In 1983 Flanagan decided to use an ellipsoid shape for his vortex generator and was able to produce a surface tension of 38 dynes/cm. By adding an ounce of this mixture to a gallon of distilled water he created a product with a surface tension of between 55 and 65 that he felt was close to the quality of Hunza water.

Surface tension (ST) is the ability of water to stick to itself, to form a sphere; the form with the least surface area for its volume; requiring the least amount of energy to maintain itself. Ordinary tap water has an ST of about 75 dynes/cm, while Hunza waters shows a much lower (less sticky) 68 dynes/cm. Washing machine detergents have an ST of about 45. Flanagan was able to produce with his 'vortex tangential amplifier' an ST of 26 (as low as ethyl alcohol), but settled on one of 38, which has greater stability.

He had notable success giving the treated water to several sick animals who quickly responded to the treatment. While strict 'double blind' tests may not yet have been attempted, these biological and structural tests have yielded impressive results.

Flanagan points out that mercury, cadmium and lead are commonly found in drinking water. These heavy metals are a major source of cancer and it is important to use a filter at home that removes them. He claims that his product is a good chelator of toxins.

Dehydration is a much greater health problem in the USA than is generally acknowledged. According to Flanagan, 75% of Americans are chronically dehydrated, due to high consumption of colas, caffeine and alcohol, climate controlled environments (both heated and air conditioned), and excessive exercise. He claims that only 2% loss of water body weight causes impaired physiological and mental performance, that mild dehydration slows the metabolism by as much as 3%, and that insufficient water is the main trigger of daytime fatigue. Other factors are confusion of thirst with hunger, and a decrease in the body's ability to make cell water as we age.

The healing power of water

Dr F. Batmanghelidj, a London-trained doctor, imprisoned in Persia (now Iran) during the 1979 revolution, was pardoned because he healed several hundreds of fellow prisoners suffering from acute, stress-induced and life-threatening peptic ulcers, by giving them glasses of tap water instead of the medication sometimes available.

Following on the success of this therapy he introduced new treatments to reduce dehydration that is often mistaken by doctors for ordinary illnesses for which they would give specific medication. Instead, he found that many of these 'illnesses' could be cured by adequate hydration. He believes that conventional medical training does not address the complex functions of water in the body. His book *Your Body's Many Cries for Water* is an international bestseller.

Batmanghelidj describes five mistaken assumptions of modern medicine. Firstly, that dry mouth is the only reliable sign of dehydration; he says that it is one of the last signs. Dehydration produces severe symptoms which medical practitioners see as diseases of toxicity requiring medication, rather than as internal, localized droughts. Another is the mistake of thinking that water's function is only life-sustaining; they dismiss its more important role of giving life.

The fourth is that the body can regulate its water intake throughout the person's life span; as we grow older we lose our perception of thirst and fail to drink adequately. Finally, many doctors believe that any fluid can replace the water needs of the body. In fact, tea, coffee, alcohol and many manufactured beverages can dehydrate and make the body more toxic.

The brain gym

An interesting discovery that was made recently is how inefficient hydration can affect brain function and integrity. It first came to light that dyslexia can be helped by drinking more water. Now many forms of autism are being ameliorated in this way. The varied forms of brain dysfunction or lack of integration are being looked at though a system known as the 'brain gym'.

Educators Paul and Gail Dennison, seeking to help children with learning disabilities, formed Brain Gym in the 1970s, building on research by developmental specialists who had been experimenting with using physical movement to enhance learning ability.

Today Brain Gym supports people of all abilities in making wide-ranging changes in their lives. Brain Gym is used in more than 80 countries and is taught in thousands of public and private schools worldwide and in corporate, performing arts, and athletic training programs. Brain Gym includes 26 easy and enjoyable targeted exercises to integrate body and mind to bring about rapid and often dramatic improvements in concentration, memory, reading, writing, organizing, listening, physical coordination, and more.

Carla Hannaford in her authoritative manual for teachers on Brain Gym, *Smart Moves,* states that our bodies are very much a part of all our learning, and learning is not an isolated 'brain' function. Every nerve and cell is a network contributing to our intelligence and our learning capability.

A key part of the Brain Gym system is drinking water before and during any mental activity to 'grease the wheels'. Drinking water is very important before any stressful situation (for instance, examinations) — as we tend to perspire under stress, and de-hydration can effect our concentration negatively. Many schools now routinely advise exam students to drink plenty of water.

Water birthing

The amniotic fluid which surrounds the foetus in the womb, is similar in composition to that of sea water. Traditionally many people living near the sea would have their mothers give birth in the ocean, which offers a natural transition for the baby. Indigenous peoples also used stream and lakes as birthing environments.

In European countries, water birthing has become relatively common. Using a portable pool became popular in the 1970s for home births. Some British hospitals also have water birth facilities. There are also agencies which will arrange a lease for a home birth. It is important to check whether your birth might have complications before committing to a water birth, which without them is usually quite safe.

The baby is given a gentle transition from the mother's womb to the mother's arms. There is no stress at all for the baby. What could be more lovely? Indeed, the mother finds a water birth gives natural pain relief as she relaxes, soothed by the warm water. There is less perineal damage. There are trained water birth assistants who may or may not be midwives.

Urine therapy

It is a recognized medical therapy in South Asia to have a patient drink their own urine to counter infections. It is a form of homeopathy, for one is taking a diluted form of one's illness to counter its symptoms. If you avoid the first flush after a night's sleep this is not as unpleasant as you might imagine! On publicizing a book on the therapy by a Dutchman who was thus cured of a serious infection in India, we discovered that it is not uncommon for Glaswegian building jocks to wash their hands in their urine when they come off the building site.[4] Apparently this keeps their hands from getting sore!

All organisms depend on water. Let's look now at the plant kingdom.

> *The amazing organizing properties of water are becoming more and more evident, which will go a long way towards explaining the detailed organization of molecules in cells and their biological functions.*
>
> Mae-Wan Ho, *SiS* New Age of Water series

8. Water Circulation in Plants

We may ask why all trees and bushes — or at least most of them — unfold a flower in a five-sided pattern, with five petals... Some botanist might well examine the sap of plants to see if any difference there corresponds to the shapes of their flowers.

Johannes Kepler (1571–1630)

The greatest miracle

For all living organisms, hydration = life; dehydration = life withdrawn.

Have you ever been in a desert environment after months of drought when there is a downpour of rain — and suddenly sprouts of new life start appearing all over the barren ground? Before long, if the rain continues, there is a verdant carpet of green. Or you may have seen this happen on a natural history movie. It happens in hot climates at the beginning of the rainy season, after months of drought.

It really is miraculous to think that there are many trillions of dormant seeds in the ground waiting — maybe for decades — for the chance to burst into life. Which ones will awaken? If they don't now, will they awake the next time? Maybe it's similar to the billions of souls said to be waiting on the eternal planes, perhaps for centuries, for a chance to incarnate into life.

What if there are the seeds of life all over the Universe, like sleeping beauties, waiting for the water prince to come and kiss them into life? There is always this potential for life, just as there is potential in all our lives for new creativity to break forth when the time and circumstances are right, or when given the right stimulus.

It is well-known amongst cactus growers that the way to stimulate flowering is to let the plant dry out, and then water it profusely.

When I was a child, there were tight flower buds made of paper (Japanese, I think) which, if you laid them on top of a water surface, would gradually open up into a beautiful petalled flower. Nature works much more magically. The desert's irrigating water carries with it information that reminds the sleeping potentiality that it can actually come to life according to its own template. The water's kiss of life is a reminder to the seed of its potential to become a plant according to universal cosmic laws. (See Chapter 13)

The more primitive life-forms, bacteria, algae and worms were early colonizers of our planet; they are still with us in huge numbers today; they built the foundations for more complex life-forms. But it was the arrival of the plant kingdom late in Earth's story that accelerated evolutionary development. They had the exceptional role of creating an oxygen-rich atmosphere and the base of the food chain that higher life-forms required.

We shall be looking at the role of water, principally with the aristocrat of the vegetable kingdom, the tree. Much is spoken today of the important role of the equatorial forests for storing CO_2, but their roles in creating beneficial climates and fresh water are seldom acknowledged and little understood.

Photosynthesis

This was undoubtedly the most important process introduced by evolving Nature, for it allowed life to take an enormous stride towards greater complexity and higher levels of energy. Photosynthesis is the process by which organisms, higher plants as well as phytoplankton, algae and bacteria, convert the Sun's energy into chemical energy. This respiration involves inhaling carbon dioxide, exhaling oxygen and storing glucose.

This marvelous alchemy transforms the basic, plentiful substance of CO_2 into essential foods. Water is the medium, or you might say, the engine, for this process. The roots bring up water replete with mineral nourishment for the plant. Photosynthesis converts CO_2 and H_2O into carbohydrates and oxygen. The breaking down of carbohydrates produces water that is transpired by the leaves which is both of greater volume and higher quality than the water that was taken up.

In this way the forests actually create water. Richard St Barbe Baker, the founder of The Men of the Trees, demonstrated that, by planting the right kind of trees, you can transform desert landscapes into productive forest that both produces new water and also attracts rainfall.

There is another, more subtle, way in which water mediates energy through trees. Life on Earth, both for its evolution and its sustenance, depends on a balance between the positive energy of the Sun and the negative energy of Earth. Water acts a bit like sperm, as the fertilizing agent in this process.[1] The tree, as the highest form of the plant, has the vital role of balancing this energy for the benefit of the whole biosphere.

Evolution of the forest

> *Only people who love it should care for the forest. Those who view the forest merely as an object of speculation do it and all other living creatures great harm, for the forest is the cradle of water. If the forest dies, then the springs will dry up, the meadows will become barren and many countries will inevitably be seized by unrest of such a kind that it will bode ill for every one of us.*
>
> Victor Schauberger[2]

Viktor Schauberger's core belief was that healthy forests are the main source of high quality water, but that they also ensure that rainfall is available in continental regions which would otherwise become arid. As equatorial deforestation has greatly accelerated since he died, it might be useful to summarize the effects of this devastation.

Plants have been around for 420 million years, which is only 9% of Earth's history. Without plants there could have been no higher life, for plants are the essential link for converting the Sun's energy into food. Trees are the highest form of the plant, and the most efficient exchangers of energy between the Earth and the Sun. The forests are the main source of oxygen, an essential building block of life; they are the Earth's 'lungs', and also vital for producing equable climates.

The establishment of forests was the essential prerequisite for the evolution of higher animals. Trees were also necessary to establish stable landscapes, allowing rivers to channel permanent watercourses and rivers, instead of the chaotic migration of alluvial flows and mud delta meanders that had been the norm (and what is found on Mars).

There have been four periods when forests have flourished: in the Carboniferous Age 350 million years ago, when land vertebrates became established; in the Jurassic, the time of the dinosaurs, 170 million years ago; in the Eocene, 60 million years ago, when primitive mammals first appeared, and in the last 500,000 years, during which the cultures of modern man developed (see chart p.50). Perhaps in each case a boost in the oxygen content of the atmosphere which the forests delivered may have been a trigger for evolutionary explosion of Earth's life-forms.

These extensive forests developed in the equatorial regions where the heat was available to prime a remarkable engine for moderating the extremes of temperature and the often-chaotic nature of the world's climates. In the first case they were evergreen forests, interspersed with enormous swamps. In the Jurassic era they were more diverse. In the Eocene when the modern great mountain ranges were being uplifted, there were large tropical jungles, perhaps not too different from the modern ones that flourished on all the continents until the late nineteenth century, but with less biodiversity.

The establishment of forests was often associated with periods of Earth's restlessness — volcanic eruptions and mountain building times, with mighty rivers providing nutrients for the plants. The forests built up a soil profile for the establishment of biodiversity, the essential conditions for evolutionary progress. Forest cover varies with climate. Forests have been the natural cover of probably three quarters of the Earth's land surface during these periods of evolutionary expansion.

Destruction of the forests

Over the possibly half a million or so years of man's time on Earth, our species has been responsible for reduction of the forest cover to about 25% of its optimum extent. The early agriculturists would burn clearings to grow their crops, and then move on to allow replenishment of fertility. Early civilizations, some well documented and some that are now folk memories, felled vast tracts of forest.

Many of these lands became desertified, becoming the Gobi, Sind, Arabian, Mesopotamian, North African and Kalahari deserts — probably through a combination of deforestation and climate change. Whole societies were uprooted and were forced to migrate in their search for subsistence. The same is likely to happen today where great swathes of the rich equatorial forests have been cleared. In those days there was somewhere else for the displaced to go, because the world's population was still relatively small. Today, however, because of overpopulation and an unsustainable birth rate, any climate changes that produce crop failures can mean only disease, starvation and the decimation of life through conflict.

Ten thousand years ago the land bordering the Mediterranean was covered with forests, mainly of oak and conifer. The forests of Lebanon provided the timber for the Phoenician empire and their exploring ships in the third century BC.[3] Two thousand years ago North Africa was so fertile that the Romans called it 'the breadbasket' of the Mediterranean; a combination of deforestation and climate change have turned it into arid desert. A thousand years ago 80% of Europe was forested; today it is about 20%, much of which is monocultured industrial woodland, which lacks the biodiversity and the energy of natural forest. In North America, before the arrival of European settlers, the forest extended from the Atlantic to beyond the Mississippi, and of course west of the Rockies.

Sometimes the forests were exploited to provide fast economic expansion, regardless of the cost to future generations (nothing has changed!). In order to provide a navy capable of ruling the seas, in the early sixteenth century, Henry VIII ordered the felling of a million mature oak trees, virtually denuding England of its finest oaks.

The proportion of the world's surface covered by forest was reduced from about 75% at its optimum to about 50% in medieval times. By 1900 it had dropped to about 35%. In the frantic rush to get rich quick, regardless of the consequences, the figure has dropped further to 25% and every year we are still losing equatorial forest the size of Belgium. It has been calculated that 20% of all global warming CO_2 emissions is contributed by destruction of the equatorial forest.[4]

Today, unstable social conditions worldwide and irresponsible political leadership favour greedy opportunists anxious to make their fortunes, usually illegally, by logging many of the finest stands of prime forests on every continent. This destruction will be seen in the future as dangerous planetary vandalism, because it will bring only more extreme weather throughout the world, loss of soils and increased desertification.

Crucially, though still little understood, forests create the environment for the propagation of water, the 'first-born' of the energies of life, as Schauberger puts it, and they moderate the climate, making it cooler in summer and warmer in winter. They are also responsible for the mineralization and fertilization of the surface soils, essential for the nutrition of higher life-forms and, most important of all, the forests create the rich humus and bacterial life, the foundation of a rich biodiversity, which stores and recycles vast amounts of rainfall, preventing floods on lower land.

Monoculture and biodiversity

If you go into a typical coniferous plantation, it is impenetrable, dark, and feels dead — a veritable green desert. No birds sing nor animals scurry, and there is little opportunity for any other plants to grow. Those that do are removed on the theory that they take away nourishment from the trees. In fact their absence increases the competition. The individual trees are all of the same age and species; they vie with each other for space and for nutrients, for all their roots go down to the same level, creating a hard pan of salts, which prevents access to the valuable minerals and energized groundwater below. There is only a certain amount of each element and chemical compound available that is suitable for that species and all the trees whose existence is wholly dependent on them must fight to get it.

It is hardly surprising that the wood from such a plantation is of very poor quality. You might compare the composition of a human community to a forest. If all the individual humans were clones of each other, and there were no elder statesmen or wise elders, how creatively barren and spiritually impoverished would be that community! These young trees are clear-felled, leaving a scene of devastation, with the valuable soil vulnerable to erosion.

A natural, undisturbed forest has a rich diversity in colour, form and vitality that bring a sense of inner tranquillity and peace. With our warped sense of order we see the profusion of life as chaotic, whereas in fact it is in the highest state of order. Order in Nature arises from a sensitive state of balance in a highly complex eco-system. What we often recognize as ordered is usually sterile and uniform. Old growth forests with mature trees with deep tap roots and minimal growth make great water catchment areas. Plantations of young, fast growing trees are thirsty and dry out the soil.

Increasing biodiversity is Nature's evolutionary imperative, for it raises the energy level, increasing complexity and intercommunication amongst species. Water is central to the quality of life. By degrading the quality of water we put at risk the health of the whole biosphere. We are now in a downward spiral of species loss and a crash in biodiversity, which it will be almost impossible to reverse.

There is a threat of worldwide famine resulting from the wheat stem rust fungus which is resistant to all fungicides. It suffers, as does the potato (viz. the Irish potato famine of the 1800s), from genetic monoculture through agronomy's plant breeding techniques. There is a similar problem with rice. The urgent need is to restore the genetic diversity of native species of all basic foods, in order to protect our future food supplies.

Tropical rainforests

The Prince of Wales has proposed a new partnership with Brazil, Indonesia and the Congo to launch better integrated rural development programmes to halt deforestation. He has emphasised the irreplaceable roles of the rainforest in providing an air conditioning system for the entire planet and for producing 20 billion tons of fresh water every day.[5]

It would cost £50 a ton to sequester carbon with new technologies being proposed. The rainforests do it free and more effectively. The Stern Report put the cost of halving deforestation at $15–20 billion. Prince Charles estimated a cost of $30 billion to stop it, less than 1% of worldwide annual insurance premiums. He is encouraging multinational businesses to commit funds to this purpose, and he has made personal approaches to heads of government with some encouraging results.

He feels the time is now right for action, as the science is clear on how little time there is left before the remaining forests disappear, with the consequent spreading of droughts, enormous loss of life and the deterioration of world climates.

The main problem is the collusion of government with international agribusiness, for ranching and to grow biofuels; and with mining and logging interests. The latest madness — the creation of enormous plantations to grow biofuel crops to satisfy modern man's insatiable dependence on the automobile — removes land from productive food production that is already down because of changes in rainfall patterns and increased costs of agriculture. The resulting food shortages cause inflated prices and immense distress in the developing countries.

The multinational companies have cynically attacked and disabled courageous campaigns by Brazilian (and international) environmentalists over several decades.

One of the richest natural experiences is to visit a tropical rainforest, for they are the priceless jewels of our ecosystem. They are vital, not just as carbon sinks or for the incredible richness and variety of their fauna and flora (Amazonia contains about 30% of all terrestrial biological material), but in substantially modifying the world's climate, making temperate regions more productive. They were on four continents, but are now only about half their extent 500 years ago: the South American is the most complete, at about 75% of its original size; the South-east Asian, from India, through Indo-China to Indonesia and Australia is about a third of what it was, and the African about 40% of its original size. The Central American has virtually disappeared.

More than twice as much of the Sun's energy reaches the Earth's surface at the tropics as in high latitudes, where the Sun's angle above the horizon is very low. The tropical rainforests of the world act as heat pumps, transferring to higher latitudes some of the enormous energy they generate, thus evening out the temperature difference. Without them, the equatorial regions would be much hotter, and the higher latitudes much colder. The larger the mass of a tropical rainforest, the more effective it was as a heat pump.

A gigantic, irreplaceable water pump, the Amazon rainforest is an essential part of the planetary circulation system, whereby a drop of water evaporated from the Atlantic is recycled six times on its way to the Andes by a process of evapo-transpiration (see Figure 14). It takes masses of humid air energy out of the Amazon basin to temperate and higher latitudes. The airflow then splits into three: the southern part is deflected as far as Patagonia; the central part flows over the Andes into the Pacific, continuing west as the trade winds; the northern airflow crosses the Caribbean, and helps to drive the Gulf Stream north-eastwards to Europe. Argentina, thousands of miles from the Amazon, gets half its rainfall from the South American jet stream, powered by the Amazon water pump. The Mid-west of the US — its golden corn belt — depends on the rain brought to it from the Amazon basin in the spring and early summer.

Rainforests act as regulators and balancers of atmospheric and oceanic systems. Now that we know, from a study of the Amazon rainforest, how the heat pump works, it is possible to conjecture that the African continent would not have been nearly as dry as it is today. In South-East Asia the destruction has reached cataclysmic proportions, with a free-for-all between corrupt local interests and greedy multi-national companies that are also extracting minerals at a fast pace, particularly in Borneo, where most of the virgin forests, theoretically protected, are likely to disappear within a decade. Brave projects are being attempted. A conservation group, trying to save the last remaining orang-utans, has in the past ten years, replanted a parcel of the cleared forest, and has managed to increase the number of primates in that region.[6]

A new theory, based on the concept of rainforests as organic rather than mechanistic thermodynamic systems, shows not only how they actually regulate the world's climates, but can also manage their own environment. Two Russian physicists, Anastassia Makarieva and Victor Gorshkov (2009), have challenged the prevailing mechanistic theory of a thermodynamic driver of air mass circulation — with a new theory of a biotic pump which is driven by the prolific tropical vegetation. The enormous area of leaf area in the forest produces a prodigious amount of evaporation and condensation — and convection which draws in saturated air from the ocean to give rise to the Trade Winds. If the natural forest is replaced by grassland or crops which cannot provide the high level of evapo-transpiration necessary to draw in the moist sea air, a reverse air flow from land to sea will dry up the soil. Without the rainforest to recycle rain, precipitation will disappear from one coast to the other, creating a desert as dry as the Negev in Israel.[7]

Makarieva and Gorshkov's thesis implies that the world cannot do without its rainforests. Instead of quibbling over how much should be conserved, we must ensure that no more forest is destroyed. Forests are not just carbon sinks or havens of biodiversity; they have an essential and irreplaceable hydrological role in the Earth's climate. They can even anticipate approaching drought by increasing evapo-transpiration through advanced leaf production.

Makarieva and Gorshkov say that biotic regulation of the water cycle also takes place in undisturbed temperate and boreal forests in the spring and summer months when the pressure gradient runs from ocean to the land. (See Bunyard 2010.)

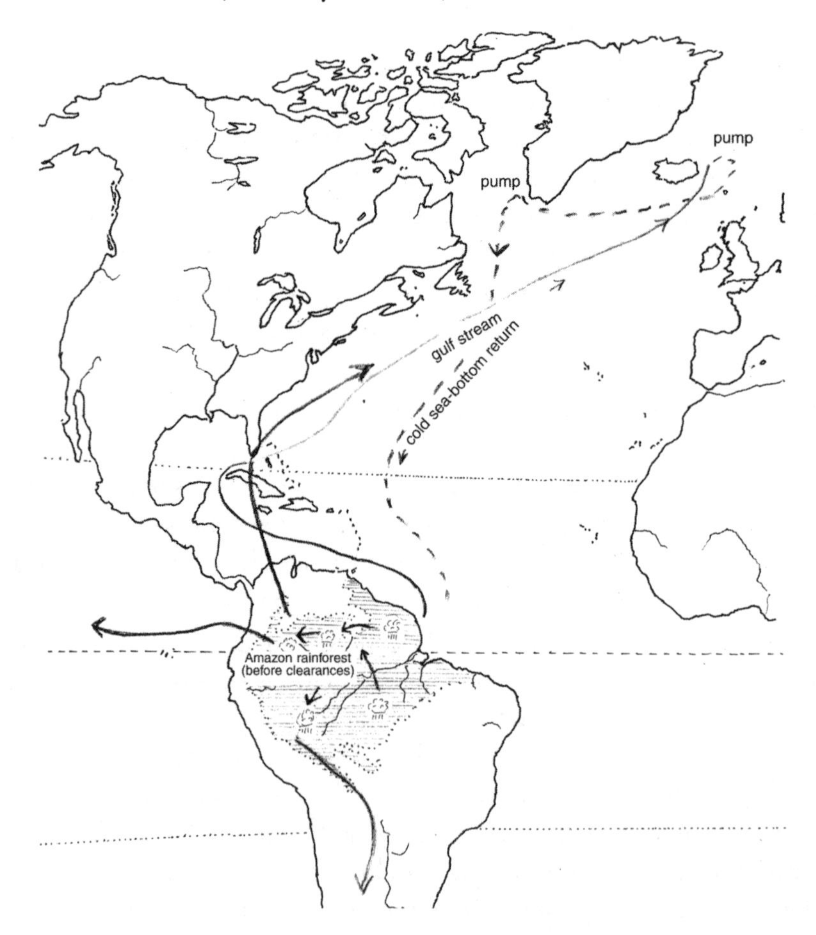

Figure 13. The Amazon heat engine. The Amazon rainforest is the engine that transfers heat from the tropics to cooler latitudes. Deforestation will reduce this vital process, causing deterioration of the world's climates. Accelerated melting of the Greenland ice-cap may cause the pumps to fail that keep the Gulf Stream flowing, causing rapid cooling of north-west Europe. (A. Bartholomew)

Temperate rainforests

Temperate forests, though a fraction of their former spread, still cover a large part of the Earth, but temperate *rainforests* occur in only a few regionswhere there is abundant rainfall precipitated by on-shore winds on coastal mountains.[8]

The most prolific is the Great Bear Rainforest which runs from Vancouver Island in British Columbia up the panhandle of Alaska, where in places the rainfall exceeds three metres a year. This magnificent virgin wilderness is richer in species than most tropical rainforests and is now the focus of an important conservation project by ForestEthics, Greenpeace, The Sierra Club of Canada and British Columbian pressure groups.

It is home to hundreds of species of mammals, including several of bear, to wolves, cougars, and particularly to various species of Pacific salmon, half a billion of which make the perilous journey every summer up the rivers of their birth to spawn. These salmon are the basic food for the bears and dozens of other animals. They do not return to the ocean like the Atlantic salmon, but die where they spawn. They have not fed in the fresh water, so they bring the energy of the ocean up to the mountain pastures. Their rotting bodies also feed the trees, providing 80% of the nitrogen which helps the Coast redwood, Douglas fir, Sitka spruce, the western hemlock and the red cedar to grow so tall and prolifically.[9] This rainforest is very fertile. Its rivers are also the cleanest in the world, as the water is filtered by the tree roots and transpired by the trees themselves in a cycle that produces the purest new water.

The creation of water

There is insufficient acknowledgment of the relationship of trees to water production. Through heir transpiration they actually create water. In an old growth forest the veterans have minimal growth, but deep tap roots which raise the water table and create a healthy water catchment area. Same age plantations in hotter countries may dry out the land but there is much evidence that appropriate species' tree planting in arid or desert conditions brings an increase in rainfall. This may be because of chemicals produced by photosynthesis, which help to generate clouds.[10] This connection has been observed in tropical rainforests, and is likely to occur in other particularly warm areas. It is one of Nature's particularly interesting feedback mechanisms.

Only when the ground surface is colder than the air (i.e. it has a positive temperature gradient) is rainwater able to penetrate the soil; (this is not generally recognized by mainstream science). The rainwater releases its free oxygen into the surrounding soil, activating micro-organisms in the soil's upper layers. Sinking deeper into the substrata, the rainwater continues to release the excess oxygen. As it cools towards the +4C° anomaly point, the remaining hydrogen can now combine with the now passive oxygen, creating new water molecules.

The maturation of water

This pure, immature water is created at the temperature when its density is highest, about +4°C *(39°F).* It sets out on a return journey from the deepest levels, becoming transformed from a hungry 'taking' substance into a mature state which is ready to nourish living systems.

The water rises through strata from which is takes different subtle energies; it becomes warmer, absorbing minerals and trace-elements. These inorganic nutrients cannot be absorbed by plants and micro-organisms, but on their upward journey the molecules become ionized; they take on an electric charge, which allows them to recombine as organic, ionized elements, which the micro-organisms and plants can absorb.

The water molecule carries right up the tree's crown the energy of the trace elements it absorbed in the roots. From the leaves' minute stomata, it is transpired into the atmosphere. On reaching its energy and temperature anomaly point at an altitude of about 3,000 metres, it is once more in a 'taking' mode, ready to absorb the finer and more spiritual energies from the Sun and the cosmos. Over time, this continuous water cycle increases the cumulative 'information' required to feed the processes that drive evolution.

During the night the descending phloem plays another important role. It interacts with suspended positively-charged xylem and because of the prevailing positive temperature gradient (see Figure 10 above) is drawn towards the exterior of the trunk. This produces new wood growth that is made denser and harder with winter cold, forming an annual ring.

In a commercial plantation a shade-demanding tree grows more branches in order to protect itself from direct sunlight. The sap is therefore diverted from its normal progress up the trunk to nourish the spurious branches, twisting around the extra knots in the trunk.

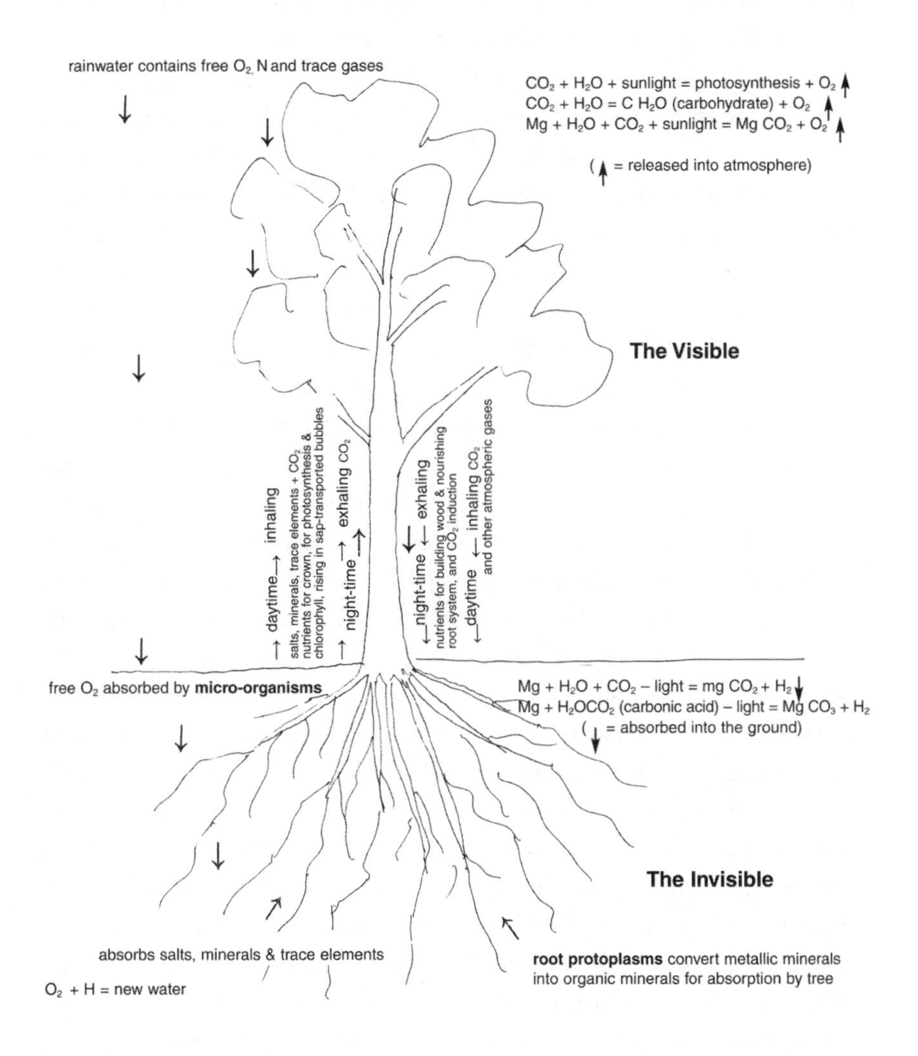

Figure 14. Tree metabolism. The vital exchange between the yang solar and the yin earth energies for the production of photosynthesis, chlorophyll and carbohydrates, and its role in the creation of water. (A. Bartholomew)

Soil and nutrition

Cooling was the key to the appearance of water, and as the ground cover spread, the lowering temperature affected the deeper ground, allowing the water table to rise, bringing minerals, trace elements and nutritional substances nearer to the surface. This created the conditions for higher quality plants to evolve. Requiring better quality nutrition, these higher plants had deeper root systems that brought up minerals from a different horizon, but were no competition for the pioneer plants.

The more evolved plants held the soil together, trapping more moisture that helped to attract micro-bacterial activity to break down the mineral particles into finer dust, the first step towards the humus that is necessary for even higher plant forms. The root systems become more complex, interweaving at different levels, so that they cannot easily be separated. Greater fertility brings a richer soil that is of too high a quality for the pioneer plants, which will now disappear. A more favourable microclimate in the higher soil increases the diversity of the bacteria, encouraging more complex root systems.

This process of soil formation took several million years before larger plants, such as small bushes and trees, were able to gain a hold; and they had to go through thousands of years of evolution before a forest could develop. The forest is the most productive environment for the building up of soil and fertile humus. It is self-fertilizing and self-sustaining. The great forests were able, over a period of thousands of years, to build up twenty feet or more of soil depth. With our heedless disrespect for Nature's bounty, in one century we have allowed these great soil banks to be eroded and destroyed, first through deforestation, and then by careless tilling of the unprotected soil surface.

The web of life that evolves in a natural forest is so complex and sensitive that the removal of key species can cause a depletion of the energy that can lead to a progressive decline of the system, as more species fail for want of the sustenance that was provided by the missing species. A hole is created in the complex root network that is the interconnecting link between the deeper ground and the surface. Because the root system raises the water table, the disappearance of a species creates a hole in the water system that supplies the nutrients. Over time, a shortage of nutrients puts more plants under stress, leading to more species disappearing.

It is clear that there is much lacking in our present understanding of the needs of plants, especially trees. Our agricultural and forestry policies are extractive rather than sustaining. Schauberger had a sensitive appreciation of the role of trees and the forest, and we would do well to take heed of his warnings before it is too late.

There is a close relationship between trees and growing food, the next stop on our journey.

> *Did the farmer know how important the forest is, he would cherish it as he would life itself.*
>
> Viktor Schauberger, *Fertile Earth.*

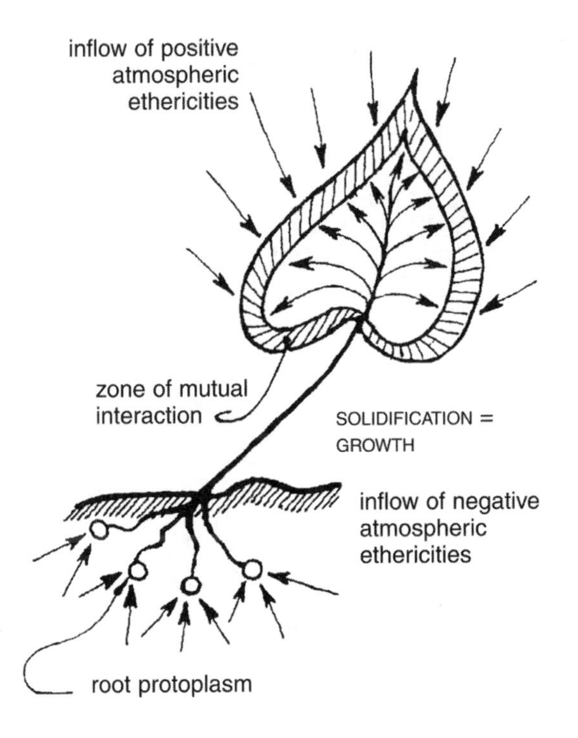

Figure 15. Energy exchange. Schauberger's diagram which shows simply how a plant is a biocondenser of positive atmospheric and negative geospheric energies. (Callum Coats)

9. The Evolution Controversy

A cell's life is controlled by the physical and energetic environment and not by its genes. Genes are simply molecular blueprints used in the construction of cells, tissues and organs.

Bruce Lipton, *The Biology of Belief*

Do you believe that the extraordinary qualities of water we have discussed (for instance, in Chapter 3), could have developed through a mechanical, accidental process? We have proposed a somewhat heretical hypothesis — that water is the creator and 'stage manager' of life, a key player in the process of evolution. It might, therefore, be pertinent to look briefly at the current debate about evolution.

Evolutionary theories

Darwin's theory of evolution rests on the assumption that life has evolved through a series of biological changes brought about by a combination of random mutations and natural selection. Those species that are able most successfully to adapt to changing conditions in their environment will supersede those less successful — 'the survival of the fittest'.

No serious student of science will deny that the evolution of species is a verifiable natural process. However, the enormous publicity given the 2009 bicentenary of Darwin's birth demonstrated that an ideology can be built around one man's work, no matter how many before and particularly after have followed similar paths of enquiry. It also shows how a pioneer's zealous followers can distort a theory, making of it almost holy writ.[1]

Over a long period, the Darwinians say, one species gradually changed into another. Fish changed into amphibians, which became reptiles; reptiles evolved into birds, which metamorphosed into animals. This theory has been taught as if it were a 'law' for well over a century (and still is), in spite of the fact that it has never been validated.

The difficulty is that the fossil record, which is the most accepted form of proof, has not revealed the transitional organisms, the intermediate forms between major groups. There is, for example, an abundance of fossils of early primates, hominids, Neanderthals and *Homo sapiens*, but no link has been found between the ape and man.

The same problem appears with plants: no intermediate fossils have been found between primitive non-flowering plants and the sudden appearance of flowering plants. Many species seem to have just appeared, without apparent links to earlier species. This is a real conundrum, a mystery. Many scientists don't like mysteries!

Faults have been found in the other cornerstone of Darwin's theory of genetic mutation. Geneticists have long accepted that mutations are usually mistakes from the DNA failing to replicate the correct information. Natural selection as a process for raising quality as well as effectiveness would need a mechanism much more reliable and predictable than genetic mutation based on chance.

There is some evidence that Charles Darwin was rather less dogmatic than many of his subsequent followers, particularly the neo-Darwinists. He was well aware of the shortcomings of his theory, calling the origin of flowering plants 'an abominable mystery'.

> The number of intermediate varieties which have formerly existed on Earth must be truly enormous. Why then is not every geological formation and every stratum full of such intermediate links? Geology assuredly does not reveal any such finely graduated organic chain; and this, perhaps is the most obvious and gravest objection which can be urged against my theory.[2]

Another problem for the Theory is that it was proposed when the prevailing theory of geological change was uniformitarianism — that geological change is very slow, a theory that, under Darwin's influence, some scholars still doggedly embrace — which would allow for gradual mutation of species. However, most geologists now accept that the big changes happen cataclysmically (the catastrophism theory), which bring with them mass extinctions of species and, afterwards, the sudden appearance of something completely new, at a higher order of complexity.

The crux of the difficulties for evolutionary theory has come with the discovery of the enormous complexity of organic processes, and particularly in the intricately linked interdependencies of biodiversity. It is hard to believe that these complexities could have evolved in a random manner or by chance; yet we still have this miracle of vibrant life working in complete synergy.

Nevertheless, Darwin's theory of evolution is a pivotal part of the current scientific orthodoxy of reductionism, which sees Nature's processes as mechanical rather than organic. A scientist who challenges it still risks his future career.

It is evident that a natural process of evolution of species does take place. Outside scientific circles, there is little awareness about the problems associated with the Darwinian theory. Many have heard about the war between Creationists and Evolutionists as a replay of the battle between religion and science, which the Scopes trial brought into sharp focus nearly a century ago.

It may not be a coincidence that the great advances in terms of evolutionary complexity coincided with the great periods of Earth restlessness — the four principal mountain-building movements. New species suddenly appeared, without precedents. How this could have occurred is an enigma. You can take your pick from competing theories: happenchance, intelligent design, divine intervention, extra-terrestrial genetic experiments or purposeful evolution.

What I am proposing here is not Creationism in disguise, but a mechanism of the intelligence of Nature in her search for greater complexity, biodiversity and 'consciousness' in life-forms. With each burst of new life-forms, a stage higher than the last is attained in terms of finer energy or consciousness. I see this form of evolution as spontaneous steps emerging out of a catastrophic situation initiated by Earth herself in the process of her evolution — sweeping away the old forms, so that something more meaningful can develop. In Chapter 12 we suggest that chaos is the prior requirement for a positive energy shift. You sometimes hear the term 'Divine Chaos' used to describe our present uncertain times. 'Chaos' is the ancient Chinese concept of the origin of creation.

Evolution needs an agent to apply the evolutionary imperative to the natural order. This agent has to be water, because I believe that only water has the sensitivity, can produce the templates, forms and patterns and can convey consciousness to organisms (see Chapter 14).

The natural history films one sees on television usually give the impression that competition between species is the engine of evolution. However, if natural systems are perceived organically, then it soon becomes apparent that cooperation between species is more of the norm than is competition. The biologist, Lynn Margulis, co-founder of the Gaia Hypothesis, insists that the survival of any environmentally inter-dependent organism shows that cooperation or symbiosis is far more important for evolution than is competition. This is more of a 'quantum' idea than a reductionist one.

The genetic challenge to neo-Darwinism

Jean Baptiste de Lamarck (1744–1829) anticipated Charles Darwin's (1809–82) theory of natural selection by fifty years. Darwin accepted Lamarck's belief that acquired characteristics could be inherited, but modern neo-Darwinists refuse to acknowledge any Larmarckian influence, insisting that natural selection is completely random in nature.

It has been noted how much scientific and religious fundamentalism have in common.[3] After James Watson, Francis Crick and Maurice Wilkins were awarded their Nobel Prize for solving the mechanism of DNA, Crick issued a 'Central Dogma' which stated that organisms are hardwired in their genetic structure, the environment and life experiences having little effect on the gene.

This is closely related to the neo-Darwinian theory which claims that natural selection favours the good genes of the powerful by making them more prolific, while the bad genes of the dispossessed are weeded out. These claims, which support the status quo and the rich, keeping the poor in their place, carry more than a hint of eugenics.

Towards the end of his life Darwin had reservations, which his followers seem to be unwilling to acknowledge, about the genes and DNA within the cells controlling our biology. He admitted:

> When I wrote the *Origin*, and for some years afterwards, I
> could find little good evidence of the direct action
> of the environment; now there is a large body of evidence.[4]

Epigenetics

The conventional concept of genetic theory, genetic determinism, holds that all characteristics are passed down by one's genes; and that we cannot pass on any influences experienced in our lifetime. That has now been challenged by the new discipline of *epi*genetics (outside genetics) which has discovered that environmental influences are more important than the genetic.

There is an urgent need to correct the mistaken dogma that our genes control our biology. Biochemistry and medicine have been more or less untouched by the revolution of the last 100 years, which has transformed the physicist's worldview from reductionism to one of interconnections. Western medicine, still dominated by deterministic principles, treats symptoms as though they were isolated from apparently unrelated parts of the organism.

The pioneers of the new biology have recognized the extraordinary damage inflicted on the long-term health of the general population as a direct result of these misunderstandings and particularly the growth of iatrogenic illnesses (now the biggest source of death in the USA). It is time for an organically based system of treatment to replace it, which would save many lives and enormous costs.[5]

The real causes of illness are social and environmental. The mainstream model of genes determining health and wellbeing is incorrect. A person's thoughts and attitude, and particularly his early upbringing play the crucial part in health.

At about the time the Human Genome Project was making headlines in the 1980s, a group of scientists initiated the new field of epigenetics which has profoundly changed our understanding of how life is controlled. It is the science of how environmental signals select, modify and regulate gene activity. In the 1990s, epigenetic research established that the DNA blueprints transmitted by our genes are not fixed at birth. The genes are constantly being modified by external influences — quality of nutrition, pollutants, social rituals, sexual cues, and by our inner environment —

emotions, biochemical and mental processes, sense of the spiritual, etc (even in the womb) without affecting the basic blueprint.

The most controversial finding was that our beliefs affect our genes, and therefore our health. We all grow up with scripts or teachings about survival. In a nurturing environment these can be good and helpful; but a dysfunctional family life can produce damaging scripts. We are all, usually as children, subjected to criticism, to self-limiting messages which may damage our prospects and our health. We are conditioned by the conventions of our upbringing and education. The hope is that at some point we may be motivated to reject our false beliefs and to reclaim our real potential.

A new therapy, Emotional Freedom Technique (EFT) has emerged in the last ten years stimulated by this epigenetic research. It works on the principle that a disruption of the body's energy system caused by trauma or by negative emotions can be corrected by tapping the ends of specific acupuncture points, with appropriate visualization. Many of these points are on water meridians, which is relevant to our theme as water is the medium of memory and association.

EFT practitioners believe that they are communicating with cellular intelligence on a holistic level rather than the more limited mental mode. It has long been recognized that the body records every experience, memory being stored in the morphic fields (see p. 212), not the brain. With the advent of epigenetics it is now clear that the cells have an overall intelligence and even wisdom that can be accessed through the water meridian.

The help of a practitioner is advised for the deeper questions, but it is a technique that one can practise on oneself, and may be used in an immediate situation, like being unable to get to sleep, or get over a block about learning a new language. It can be effective with conditions that may not seem to have an emotional trigger (for instance, arterial sclerosis). It has implications for radical understanding of the individual human dilemma that most therapies don't easily reach.[6] We *can* change our lives through having a positive attitude and thoughts!

Epigenetics has created the most profound shift in thinking about the human predicament. The old biology sees the human individual a victim limited by his/her genetic situation. The new biology emphasises self-empowerment and the opportunities to optimize one's innate gifts, validating principles of the human potential movement of the 1970s.

The cells

Single-celled organisms appeared six hundred million years after Earth's formation. Two and three quarter billion years later they evolved into multi-cellular organisms which would eventually contain up to trillions in number.

The cell is an intelligent being that can survive on its own. Bruce Lipton claims that they show intention and purpose when they actively seek environments that support their survival, and avoid hostile ones.[7] They are capable of learning through these environmental experiences and of creating an antibody blueprint for a specific virus (for instance, measles). They pass on their environmental experience to their offspring who retain the genetic 'memory' of its antibody protein developed to cope with an invading virus.

Cell biology is one of the best examples of the holistic principle of fractals, one of the principal techniques for evolutionary advance. A fractal is a design or pattern repeating itself at different magnitudes (see p. 213). The 'primitive' single cell organism was the template for the evolution of highly complex evolved organisms of trillions of cells, only there is specialization of function to optimize efficiency and survival.

The more that cells are in touch with their environment, the better the survival chances of an organism. Epigenetics shows that the genes are the physical (biochemical) memory of an organism's learned experience. In the New Biology, evolution becomes survival of the fittest *group*, not the individual (cooperation, not competition). While DNA is the blueprint of a person's potentialities (like a birth chart?), the genes are the material the organism has to work with, which are modified by experience.[8]

Even though humans are made up of trillions of cells, there is no function in our bodies that is not already expressed in a single cell, which has the functional equivalent of nervous, digestive, respiratory, excretory, endocrine, muscle and skeletal, circulatory, skin, reproductive and even immune systems.

Epigenetics recognizes two mechanisms by which organisms pass on hereditary information: nature (through the genes) and nurture (epigenetic). If you focus only on blueprints (the old biology) environments seem totally irrelevant.

The old biology

—The mechanisms of our physical body can be understood by dissecting cells down to their building blocks (reductionism).

—The linear flow of information is one-way — from DNA (long-term memory) to RNA (template for synthesizing protein) to protein.

—Our genes are fixed — we are lucky to have good genes, and unlucky if we don't. It's a chancy business! All characteristics are passed down by our genes; we cannot pass on any influences experienced in our lifetime.

—Watson and Crick proposed that DNA controls its own replication and serves as the blueprint for the body's proteins. DNA 'rules', according to the Central Dogma.

—The primacy of DNA provides the logic for the Age of Genetic Determinism.

—The Human Genome Project spent billions on the assumption that our biology is controlled by our genes. Only 25,000 genes were discovered — one sixth of what was expected, no more than a humble worm possesses (it was a very poor investment).

The new biology

—The New Biology emphasizes that coherence on all levels, cellular, molecular, atomic, and organic governs all life processes.

—Quantum physics with its view of the interconnections between all life forms now demonstrates that our DNA is controlled from signals outside the cells, including our personal 'scripts', messages from positive and negative thoughts, from the environment and from the experiences of the whole organism.

—The chief implication of this is that, as organisms, we can choose to influence our own evolution because our thoughts and experiences strongly influence our futures. 'Mind Over Matter' is a reality! People's lives are frequently stunted or destroyed through holding onto false or disempowering beliefs. Epigenetics is often called 'The Science of Self-Empowerment'.

—The flow of information is multi-directional — a complex network of paths created by resonance interactions (each on its own wavelength). The New Biology is superbly holistic.

—To quote Bruce Lipton: 'Biomedicine doesn't recognize the massive complexity of inter-communication between physical parts and the energy field that make up the whole. Cellular constituents are woven into a complex web of crosstalk, feedback and feed-forward communication loops. A biological dysfunction may arise from a miscommunication along any of the routes of information flow.' (that is, the cause of iatrogenic illness)[9]

The cell membrane (1 millionth of a millimetre thick, and discovered in the 1950s with the electron microscope), a three layered skin holding cytoplasm together, is the cell's 'brain' (homologue of a silicon chip with its memory — 'a liquid crystal semi-conductor with gates and channels'). The nucleus is not the cell's brain (old biology).

Embedded in the hydrophobic middle layer are: receptors — the equivalent of sensory nerves — which monitor specific external and internal signal; and effectors — the equivalent of action-generating motor nerves, resonating to specific vibrational frequencies or by shape and electric charge locking on to a histamine molecule, for example. These receptors are called IMPs (integral membrane proteins). Like a modern computer, the IMPs are programmable from outside — effectively a biocomputer.

I believe this is a similar water signal mechanism to what we suggested was working with Cleve Backster's biocommunication experiments (see Chapter 14). Mae-Wan Ho points out that this is because of the innate coherence of the organism, both within and without, which makes it responsive to everything around it.[10]

It's not the quality of the human genetic makeup, but the experience of the whole organism that can then be passed on to successive generations, usually selective within the same sex. In another epigenetics study, ground-breaking research at the Institute of Child Health, University College, London, demonstrated how young boys' experiences could affect not only their own health in later life but also the health of their sons and grandsons.[11]

'Intelligent design' or 'purposeful evolution'?

The biblical Creationists' simplistic belief in the origins of life discourages credibility. The 'Intelligent Design' theory (a new trend in American Christian education) that has Creationist roots, says 'life somehow assembled itself out of organic molecules'.

Another theory, which has yet to be elaborated, is that knowledge of edible grains was somehow passed from an earlier, developed but forgotten civilization. A more recent theory claiming attention, the 'Interventionist', insists that life-forms were introduced by extraterrestrial civilizations; needless to say, this theory has no orthodox support, being regarded as too fantastical.

Linked to the Catastrophism position — and perhaps the most interesting — is the theory of Continuous Creation, which is in direct opposition to the now generally accepted Big Bang theory that satisfies religious groups who believe in a single primordial act of creation by God. This new theory is proposed by Paul LaViolette who believes that the purpose of the relatively rare cosmic event of the galactic super-wave, caused by massive explosions at the galactic core, is to create matter from the 'etheric flux' that invisibly pervades the entire Universe.[12]

The last occurrence may have been about fifteen thousand years ago. The Vostock ice cores in Antarctica show a peak of cosmic radiation and a sharp increase in temperature at this time. LaViolette claims that this actual event could account for the classical Greek writer Ovid's description of a scorched Earth phenomenon and also for some eighty different indigenous societies' flood myths.

Why don't the more realistic theories of evolution get a look in? Probably because the emotional polarization between neo-Darwinism and Creationism creates a din that drowns out the more moderate concepts in between. The way people hold to one point of view or another is baffling. This stubbornness is both unscientific and irrational. There is much evidence for evolution, and the possibility of a creation plan is also hard to completely discount. What is so wrong about being open-minded and allowing room for both? Taking a dogmatic position does not help find the truth.

The concept of an etheric substratum from which matter is created, that originated in Hindu metaphysics, has attracted considerable scientific credence over the years. It was more recently revived by David Bohm, who saw the Universe as part of something far vaster, more ineffable and essentially conscious.

My proposal holds a similar view of consciousness at all levels of life, but seen at the Earth rather than the Cosmic level, as a mechanism of intelligent Nature searching for greater complexity, biodiversity and 'consciousness' in life-forms. The quantum or etheric field may be seen as the ground of consciousness, water as the vehicle for transmission and communication.

The evidence shows that, with each burst of new life-forms, a stage higher than the last was attained in terms of complexity and biodiversity. The great evolutionary advances always seemed to follow cataclysms of some kind. I see this form of evolution as spontaneous steps emerging out of a catastrophic situation initiated by Earth herself in the process of her evolution, a theory quite dissimilar to any of those described above. We might call it 'purposeful evolution'.

Consciousness, meaning and coherence

Some may object to the concept of consciousness being introduced into evolutionary theory, but I feel it's reasonable as part of the increased complexity and inter-connectedness that evolution presents — consciousness defined as a level of perception of relatedness on a hierarchical scale. It is the companion of spiritual meaning favoured by those who take the wider worldview. Some of the scientific pioneers, like Albert Einstein, Sir James Jeans or Carl Gustav Jung espoused a spiritual worldview.

Dr Mae-Wan Ho equates the idea of organic coherence with the notion of consciousness:

> Quantum coherent organisms invariably become entangled with one another. A quantum world is a world of universal mutual entanglement, the prerequisite for universal love and ethics. Because we are all entangled, and each being is implicit in every other, the best way to benefit oneself is to benefit the other. That's why we can really love our neighbour as ourselves. It is heartfelt and sincere. We are ethical and care about our neighbours and all of creation because they are literally as dear to us as our own self.[13]

A teleological discussion

In recent years there has been a rapprochement between esoteric theories of the evolution of life and some more holistic scientific ideas. Quantum physics now entertains the possibility of unifying concepts, suggesting even a sense of purpose in evolution. Creation myths and Eastern esotericism have traditionally regarded the Earth as empowered with a level of intelligence, which has been mirrored by contemporary scientists like James Lovelock and Lynn Margulis. Their Gaia hypothesis (1979) holds that the biosphere is a self-regulating entity that keeps the environment constant and comfortable for life. (As a geochemist, Lovelock's principal concern at the time was the constitution and homoeostasis of the atmosphere, rather than the origin and evolution of life itself.)[14]

Taking the self-regulating concept of Earth a step further brings up the theory of Earth as an intelligent organism on its own evolutionary path, with Nature's evolution as a dependent part of the system. At this point there enters a spiritual dimension. In esoteric and spiritual traditions Nature is called 'the mirror of the Divine', which could also be identified as a level of 'being' of a more evolved Earth — manifesting complexities of life-forms in response to Earth's own evolution. The evolutionary imperative of Nature, in these traditions, seems to be towards greater complexity and biodiversity, and towards higher levels of purpose.

Water and evolution

We have seen how evolution and biodiversity are dependent on water, and how water's complexity changes with the demand of the evolutionary blueprints. In the last chapter we saw how water is responsible for driving climate change. It is clear that the future prospects for life on this planet will depend on water. We would do well to try to communicate with water's consciousness, for this might give us clues about our options for the future.

We have, up to this point, been reviewing the role of water as it is generally understood by mainstream science. In the next chapter we shall consider the contribution that Viktor Schauberger, a pioneer of 'the Science of Nature', has made in opening up science to a more holistic view of how Nature works, for in Part 2 we shall be discussing the new insights about water as the source of life.

> *The majority believes that everything hard to comprehend must be very profound. This is incorrect. What is hard to understand is what is immature, unclear and often false. The highest wisdom is simple and passes through the brain directly into the heart.*
>
> Viktor Schauberger

10. The Water Wizard

Water in the environment is like blood in the body, and ours is sick. The arteries and veins of our countryside, its rivers and wet-lands, are suffering from the equivalent of low blood pressure and blood poisoning. The condition has developed over many years and treatment is now urgent.
Sir David Attenborough

We cannot progress in our study of water without introducing the groundbreaking research of the man who was known as 'The Water Wizard'. Viktor Schauberger (1885–1958) was a genius whose ideas were way ahead of his time. He challenged conventional thinking about natural processes and, probably more than anyone else his extraordinary insights into water have compelled us to think 'outside the box'.

What was extraordinary about Schauberger was his ability to live in two separate worlds. He was at heart an intuitive, practical man with inherent engineering skills, from a family with a long tradition of caring for the forest. But he could see, as well, how modern education was stifling the psychic sensitivity that enabled his predecessors to understand how to work with Nature. So, in reality he was an odd combination of an ingenious inventor and an intuitive shaman. He was a scholar as well, a frustration to his wife with his habit of staying up late at night studying esoteric texts.

As a young man he worked as a 'forest master' in the Austrian Alps when they were still a true wilderness. This experience was to influence his entire life's work. From his precise observations of natural phenomena, Schauberger pioneered a completely new understanding of the nature of water as the most important life-giving and energy empowering substance on the planet.

Viktor Schauberger's insights into water

Schauberger intuited both the nature of living water and its importance for life processes, but also the natural conditions required for its manifestation. He was a holistic scientist who saw water as a living organism. The necessary conditions were largely to do with how water wants to move in a spiralling, centripetal (in-winding) manner which cools it and increases its density, encouraging the flourishing of life-forms.[1]

Schauberger saw water as a pulsating, living substance that energizes all of life, both organic and inorganic. Whether as water, blood or sap (which are essentially water), it is the indispensable constituent of all life-forms, and its quality and temperature are fundamental to health. When it is healthy it has a complex structure that enables it to communicate information, carry subtle energy, nutrients and healing, to self-cleanse and discharge wastes.

He believed that one of the causes of the disintegration of our culture is our disrespect for and destruction of water, the bringer of life, for in doing so we destroy life itself. Viktor also profoundly believed that our inappropriate technologies deliver to our homes poor water that has lost its dynamic energy and its ability to pulsate — and is effectively lifeless. This dead water delivers inadequate nutrition, and Viktor believed that its regressive energies are responsible for degenerative diseases like cancer, for lowering human intelligence and for communal disharmony.

Natural forests (not the monoculture plantations of today) are the cradle of water and also the main source of oxygen for the planet. Their precipitate destruction, Schauberger predicted, would result in climate change, severe water shortage and the creation of deserts. He made brilliant observations of the way in which trees in a natural, diversified environment create water, are biocondensers of subtle energy (accumulating, storing and exchanging energy from both Sun and Earth) — and how the groundwater brings Earth's energy to the tree in order to balance its intake of the Sun's energy.

The most radical of Schauberger's discoveries are:

Motion

He discovered that all life and growth are shaped and fashioned through vibration and movement. There are two forms of flow or motion in Nature: the outward moving, expanding, (centrifugal), which Nature uses to break down and decompose, and the inward moving, contracting (centripetal), which Nature uses to build up and energize. Our current technology uses the 'wrong' form of motion!

Our machines and processes channel agents such as air, water, other liquids and gases into negative centrifugal motion. As a consequence, these substances are devitalized and debilitated, which affects their surroundings. The dynamic energy produced by our technology is harmful because, by its very nature, it causes deterioration in the environment by strengthening those subtle energies that break down structures and degrade quality, while at the same time suppressing those that increase the quality of the dynamic energies that help plants and animals to be healthy.

This can dangerously affect the vital biodiversity and balance in our ecosystems. Our mechanical, technological systems of motion are nearly all heat- and friction-inducing, which makes them noisy and inefficient; this is how we generate our power — contributing significantly to global warming through entropy (waste heat).

Nature uses the opposite, centripetal form of motion (moving from the outside inwards with increasing velocity), which acts to cool, to condense, to structure, assisting the emergence of living systems of higher quality and greater complexity. Using his knowledge of the way that Nature energizes and purifies water, Schauberger was able to develop a spring water machine that transformed poor quality water into drinking quality.

Most of us are aware of the polluting effects of chemicals in the body and in the soil, of the dangers of radioactive waste and the risks of biotechnology. But Schauberger was also concerned with something much more basically wrong with our technology. As a practical man, he observed the appalling squandering of resources; why are the internal combustion and steam engines on which our civilization depend not even 50% efficient? The energy that is not turned into power or motion is wasted and heats up the atmosphere, adding to the greenhouse effect. From his observations of Nature came the answer — that we use the wrong form of motion.

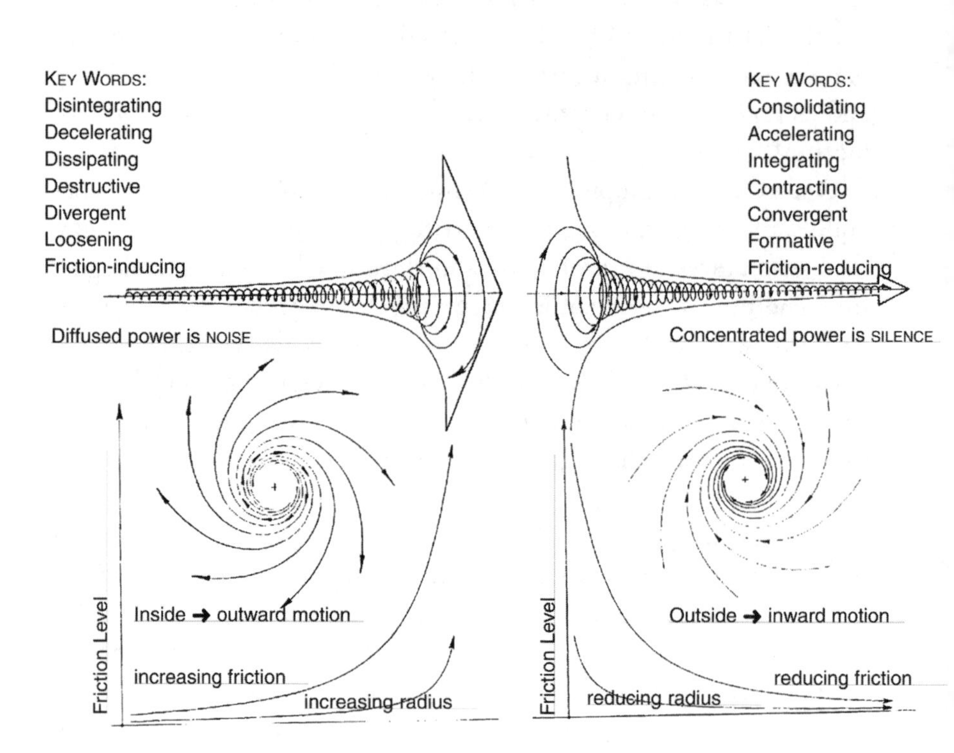

Figure 16. Centrifugal and centripetal movement. On the left, how our current technology works (axial>radial — inside>outwards). On the right, Nature's way of generating energy (radial>axial — outside>inwards). (Callum Coats)

Temperature

We are familiar with the principle that the normal temperature of blood in the human body is 37°C *(98·4°F)*. A very small change in that temperature indicates sickness. It is the same with water and with sap. One of Viktor Schauberger's most important discoveries was to do with temperature. He showed how small variations of temperature are as crucial to the healthy movement of water and sap as they are for the human blood. Heat always moves towards cold. We know that humans must live within a certain temperature range in order to survive, but few realize how narrow this is for all living things.

Schauberger insisted that temperature change is the most important catalyst in Nature. He identified two modes — a rising gradient to break down and decompose (he called this the 'negative' gradient), and a falling gradient (approaching the anomaly point of 4°C *(39.2°F)* to build (the positive gradient).

Having learned from his family about the importance of water temperature, Schauberger decided to do a demonstration for the world-renowned hydraulicist, Professor Philippe Forchheimer. He had colleagues heat up 100 litres of water that they poured into the large stream, some distance above a stretch of rapids, 500 metres higher than where he stood. Viktor observed how the trout they had been observing became agitated, and soon was unable to hold its station in the fast flowing stream, being swept down the rapids. Forchheimer was astounded; mainstream science does not yet accept that such small changes in temperature could make such a difference to the water environment.

The minute rise in the average temperature of the water had interfered with the trout's hovering ability. Viktor searched the textbooks in vain for an explanation of this strange phenomenon. The marginal rise in temperature upstream caused a loss of coherence in the water, so that the trout was unable to hold its station, and was swept downstream.

In the natural process of synthesis and decomposition in all waters, trees and other living organisms, both the rising and falling temperature gradients are active. Each form of gradient has its special function in Nature's great production; the positive (cooling) temperature gradient must play the principal role if evolution is to unfold creatively.

This important factor affects all the features of a river, such as flow velocity, tractive (pulling) force, sediment load, turbidity and viscosity, and everything to do with water management generally, like its storage and transport through pipes. It is because modern hydrologists do not recognize the temperature gradient that they are unable to prevent rivers flooding or to deliver better quality water to our homes.

Flowing water behaves according to whichever temperature gradient is active. The positive temperature gradient builds up living systems by cooling, concentrating, and energizing as it approaches +4°C *(39°F)*. The key to this process of healthy growth and development is that the ionized substances are drawn together into intimate and productive contact, and the contained oxygen becomes passive and is easily bound by the cool carbones, the building blocks of life.[2] The increasing warming of the negative temperature gradient, however, reduces the cohering energy and loosens the structure of an organism and the forms start disintegrating. The oxygen becomes increasingly aggressive and instead of helping to build structures, pulls them apart, risking pathogenic disease.

Polarity

Nature is founded far more on cooperation than on competition, because it is only through harmonious interplay that physical formation can occur and structures can be built up. At the heart of the creative process in Nature are polarities such as positive and negative, chaos and order, quantity and quality, gravitation and levitation, electricity and magnetism. In every case, for any natural process to be harmonious, one polarity cannot be present without the other, and each needs the other to make up the whole. Natural law, as Schauberger describes, requires that, for creative evolution to be maintained, the polarities are not 50/50, which would result in atrophy, but are unevenly balanced towards the *yin* or negative.

Schauberger studied biomagnetism and bioelectricity which are two complementary qualities. These always operate simultaneously because every thing is bi-polar. In every man a woman exists, and in every woman a man. The catalytic role of dual polarity starts with the positive charge of the Sun and the negative of Earth, but is an essential component for all biological processes.

The memory of water

Water's reputation as a powerful solvent derives from its electromagnetic qualities. The positive hydrogen atoms in the water molecule attract to themselves negative ions from the substance they are in contact with, while the oxygen atom with a double negative charge joins up with positive ions, so that balance is maintained. In this way water breaks down and dissolves substances into their constituent parts, taking oxygen, nitrogen and carbon dioxide from the air, and calcium, potassium, sodium, manganese, and so on, from the rocks. Water continually collects substances from one source, depositing them, usually as building blocks for new growth, somewhere else.

When water is flowing as its nature dictates, energetically in spirals and vortices, it creates the structure necessary for it to carry constructive information. This structure comprises microclusters of vibrating energy centres, constantly receiving and transmuting energy from every contact the water body makes. Despite water's fluidity and its ability constantly to change its state, the molecules, if conditions permit, generally organize themselves into structures or clusters.

The clusters can store vibrational impressions or imprints. If these are beneficial, they may be able to restore healthy resonance in the human body, as through homeopathy. On the other hand if they are the imprints of toxins or pollutants in the drinking water, they may be carriers of disharmony and disease.

Viktor Schauberger demonstrated that water as an organism has a life cycle from birth through maturation to death. When it is treated with disrespect or ignorant handling, instead of bringing life and vitality, it becomes anti-life, facilitating pathogenic processes in the organisms it inhabits, which initiate physical decay and eventually bring death. One of Schauberger's more controversial discoveries was that water that has lost its coherence takes on negative energy that precipitates deterioration in the human being, affecting our actual moral, mental and spiritual wellbeing.

If science were able to see water as possessing as well as giving life, it would be a giant step towards the rehabilitation of water in human society.

The symbol H_2O represents pure or distilled water. Schauberger called this water juvenile, because it has no developed character or qualities. When it is immature, water takes, absorbing minerals with a voracious appetite, to give back the much-needed nourishment to its environment only when mature as a mountain spring. Water is mature when it is suitably enriched with raw material, what we call 'impurities', on which other organisms depend for their energy and life.

Water has a memory; when we think we have 'purified' water of the chemicals and hormones we have mindlessly thrown in, in order to make it drinkable, the energy of these contaminants remain, polluting our energy bodies in the same way that chemicals affect our physical bodies. Because of its nature, water sacrifices itself entirely to the environment, for good or for bad.

There was little support for Schauberger's insights about water, and it was not until the development of chaos theory in the late 1970s that hydrology, together with the discoveries of quantum mechanics, started to catch up with Schauberger's research.

> *Everything flows, floats and moves. There is no state of equilibrium — there is no state of rest.*
>
> Viktor Schauberger, *Nature as Teacher*

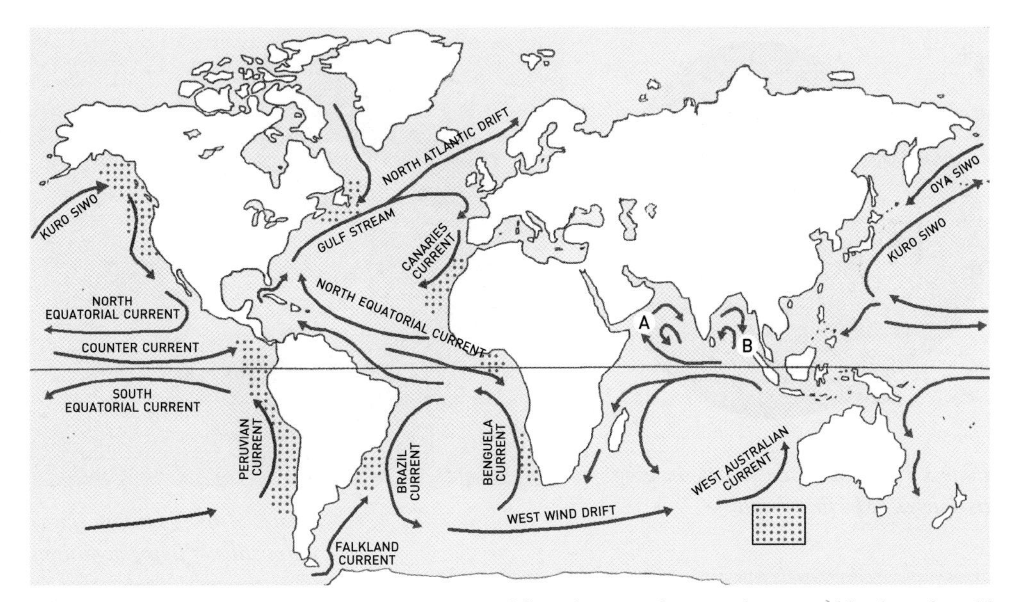

Plate 1. Ocean Currents: The main ocean currents follow the prevailing wind systems: blue lines for cold currents, black for warm. The reversal of the monsoon wind system causes a reversal of ocean currents in the Arabian Sea (A) and the Bay of Bengal (B). Upwelling water brings up nutrients that support rich breeding areas for fish and cetaceans. (R.B. Bunnett)

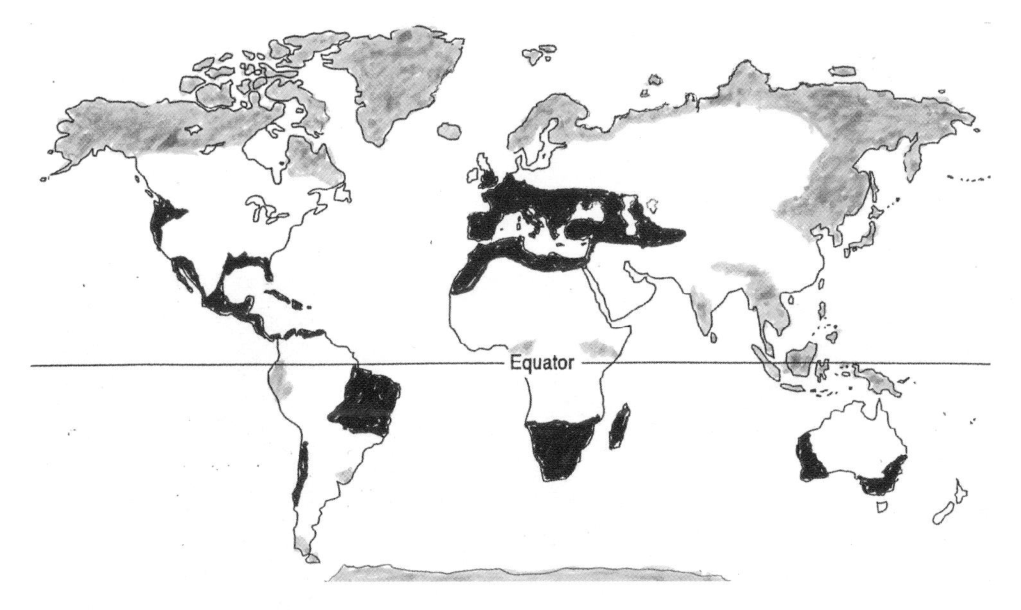

Plate 2. Regions of Water Scarcity c. 2090: Areas of drought (red) are predicted for southern Europe, the Mediterranean and the Middle East; North-west and South-west USA, the Caribbean (surprisingly), South Africa and Brazil. While increased precipitation (blue) is agreed for China, equatorial areas, North Polar regions and below 60° S latitude. (IPCC scenario 2007, changes relative to 1980-99) (A. Bartholomew)

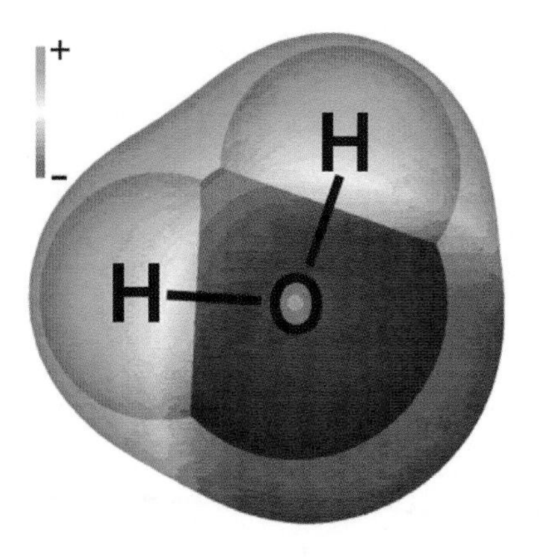

Plate 3. Water Molecule Model, showing its charges. (Martin Chaplin)

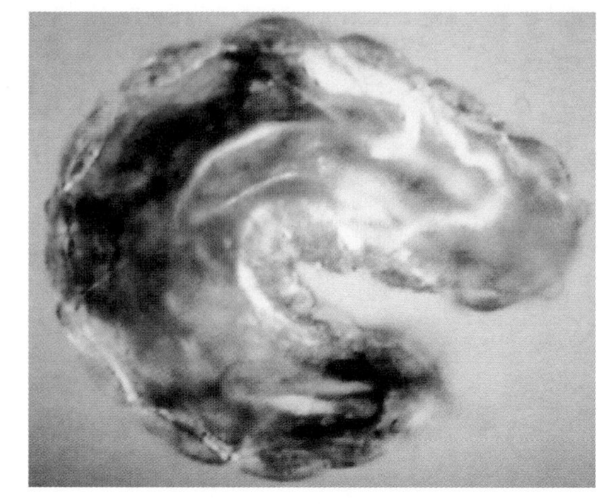

Plates 4-6. Quantum Jazz: Film stills of living organisms seen through a polarizing (x 40) microscope, their pulsating energy shown in vivid colours: Twisting worm; Daphnia; Artemia. (Institute of Science in Society)

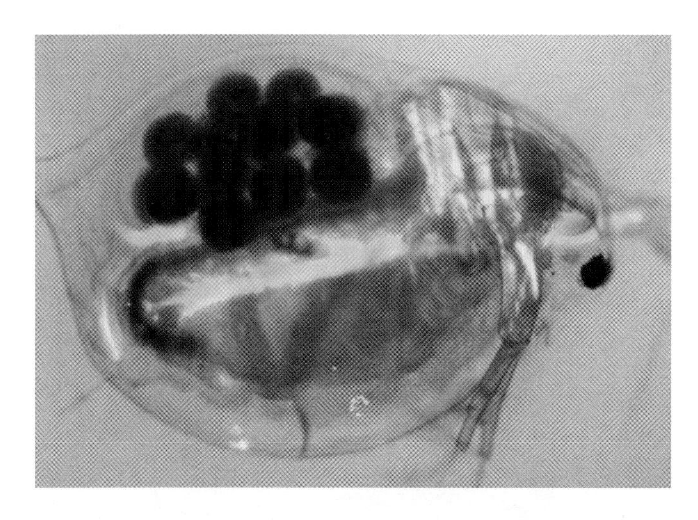

Plate 7. Double Egg Vortexer (developed by Ralf Roessner). This demonstrates how the orientation of an egg affects the treatment of water: yang (expansion in top egg) and yin (contraction in the lower). (A. Bartholomew)

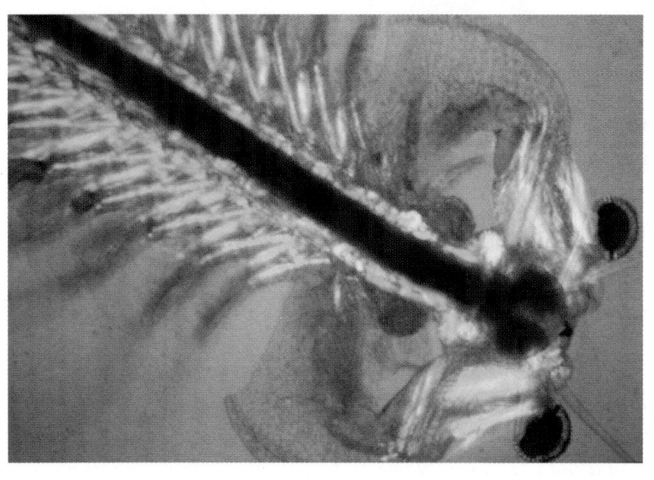

Plate 8. Surfer going 'down the tube'. (Russell Ord)

Plate 9. Järna Flowform. (J. Wilkes)

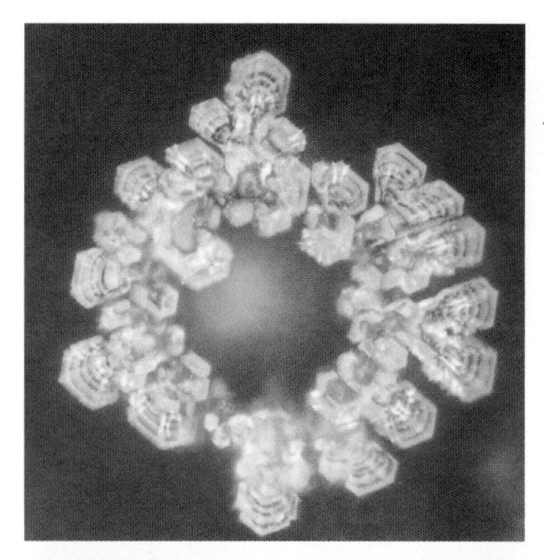

Plates 10-14. Emoto's ice crystals, frozen samples of water subjected to energy either at source or from words attached to sample bottle. (Masaru Emoto)

10. Lourdes Spring

11. Fujiwara Dam, before a healing prayer

12. Fujiwara Dam, after a healing prayer

13. Bach's Air on a G String

14. Heavy-metal music

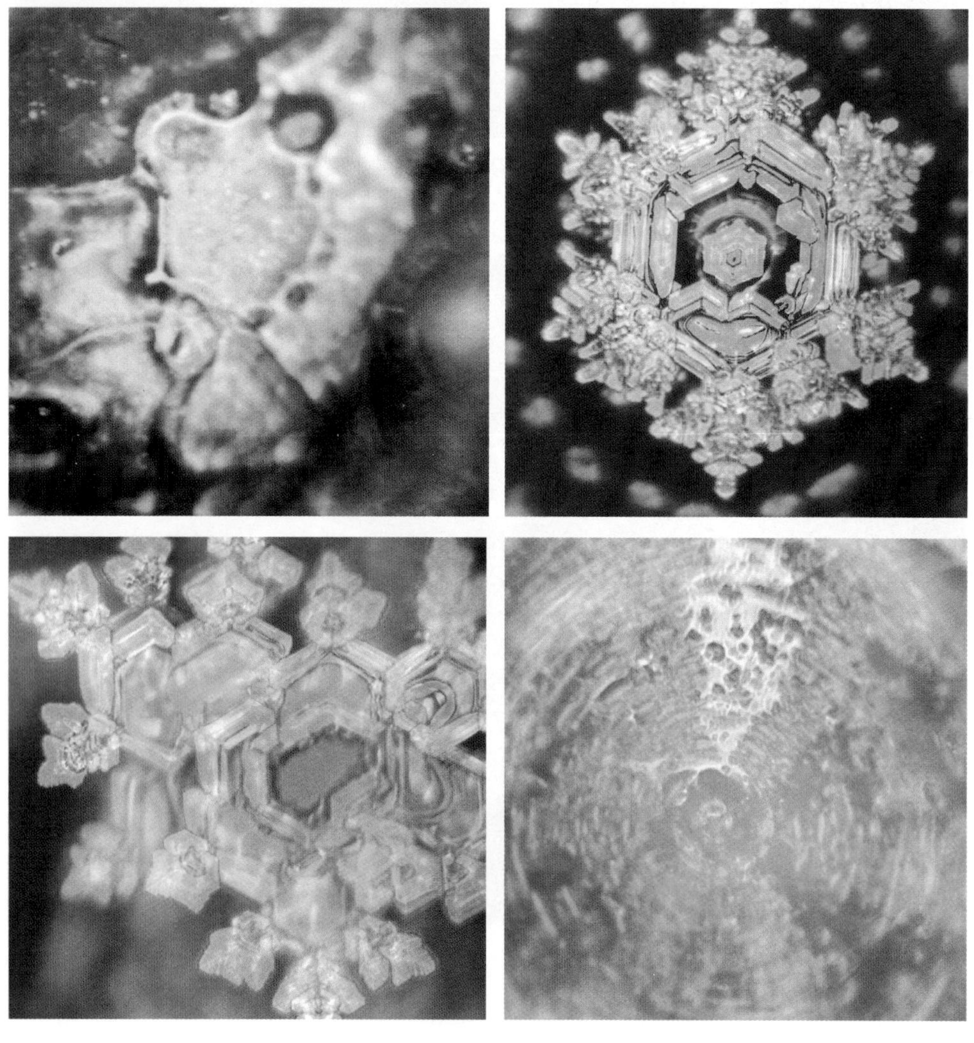

15. A Dartmoor spring (A. Bartholomew)

16. The vortex of a whirlpool. (Pythagoras-Kepler School)

17. Sacred geometry - plants reflecting geometrical laws in a calabrese. (A. Bartholomew)

18. The aurora borealis, with its dynamic motion, behaves like water does on Earth, often forming spirals and whirlpools. (from Falck-Ytter, Aurora)

Plates 19-21. David Schweitzer's dark field microscope photos: These were taken through a polarizing filter at a magnification of x 4,000. Tap water was treated with a vortex energizer, and two areas of the sample showed different structures. The dendritic form is thought to represent what Schauberger called the female or yin aspect of water and the crystalline formation the male or yang. (Centre for Implosion Research)

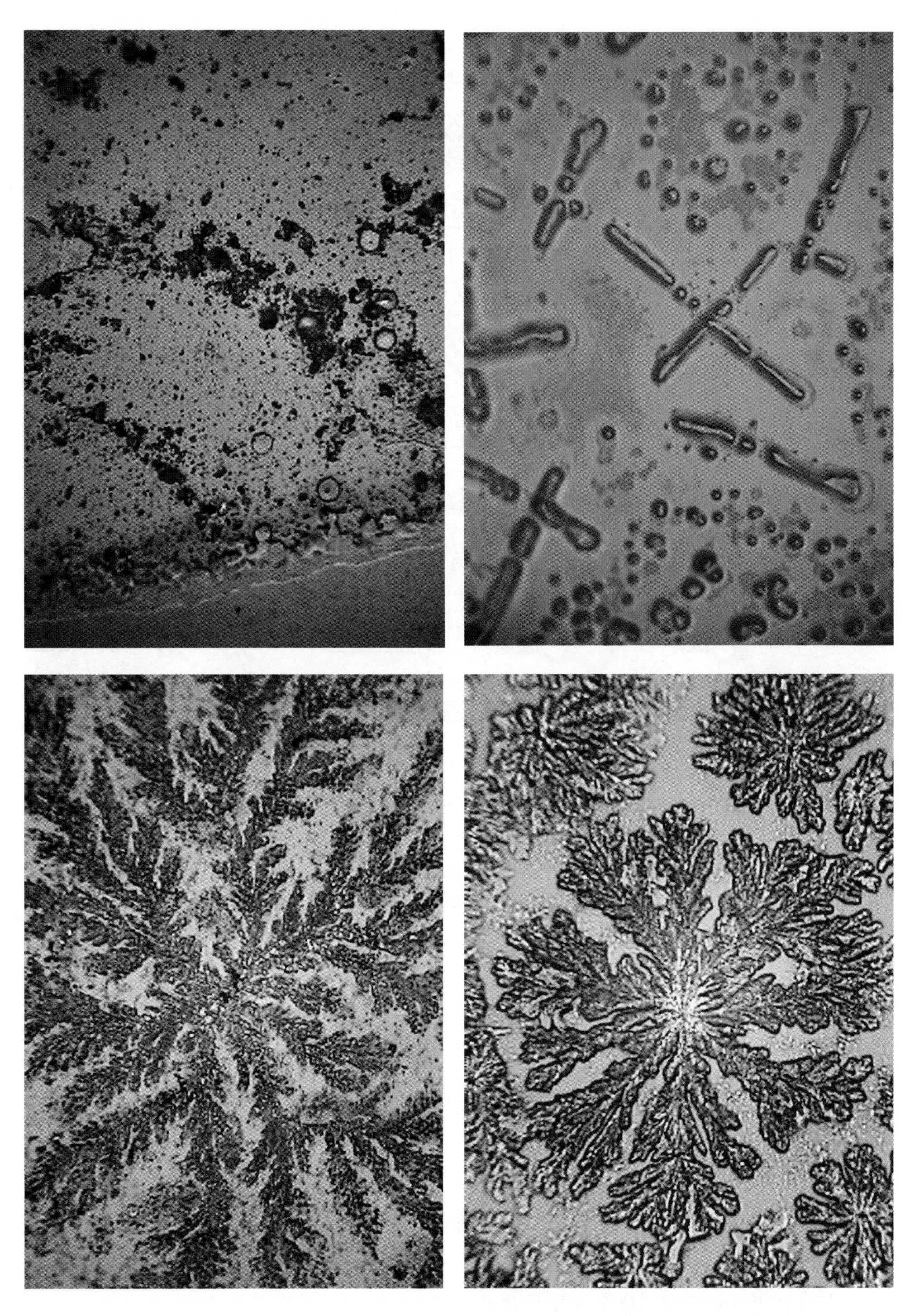

Plates 22-25. Andreas Schulz' Water Analysis Crystals: Analysis based on the angular relationship between crystals formed from a combination of distilled and desiccated samples. 22. Paris tap water; 23. Stuttgart mains water (c.300x); 24. Spring water from New Zealand (c.100x); 25. Spring water from La Palma (c.300x). (Andreas Schulz, Water Crystals)

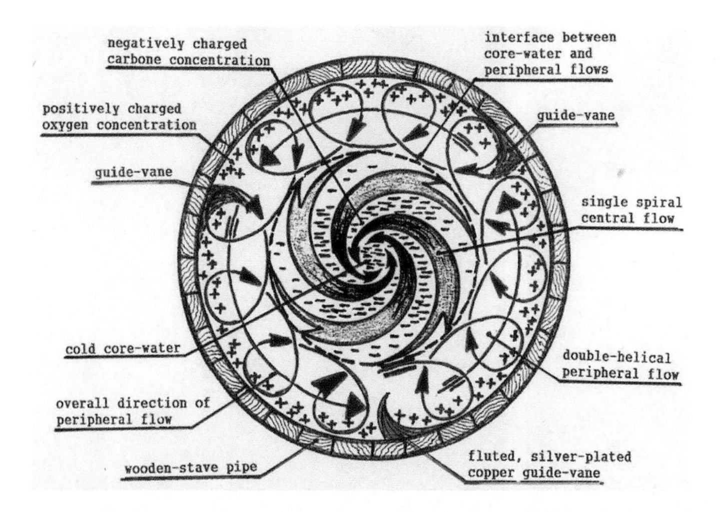

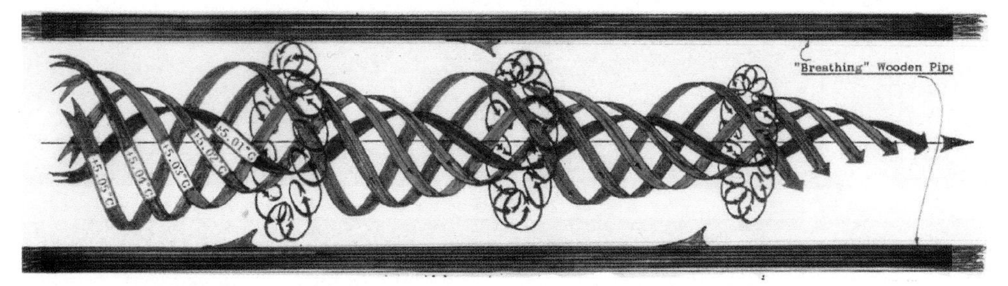

Plates 26-27. Schauberger's wooden pipe. (Callum Coats, Living Energies).

26. Cross-section: The action of a torroidal vortex, where the concentration of positive oxygen destroys the pathogenic bacteria, acting like the river's immune system.

27. Longitudinal section: Guide vanes create torroidal counter-vortices to transfer impurities to the pipe walls where the oxygen concentration destroys the pathogenic bacteria. They also act like ball bearings to enhance the forward movement

Plate 28. A theoretical super cluster: thirteen interpenetrating icosahedra, showing how they easily self-replicate like fractals. (Martin Chaplin)

PART 2

Water as the Source of Life

11. The Organism and Quantum Water

I think we are beginning to perceive nature in Earth in exactly the opposite way we viewed it in classical physics. We no longer conceive of nature as a passive object . . . . I see us as nearer to a Taoist view, in which we are embedded in a universe that is not foreign to us.
Ilya Prigogine

The quantum field and the ether

A few of the early pioneers of quantum physics in the early twentieth century sensed that their discoveries would revolutionize biology, cosmology and concepts of life, but it has taken a new generation after World War II to discover that quanta fill the macro environment as well as the microscopic, making an enormous web of interconnected dynamic energy that seems to continue infinitely through space, a kind of communication system.

However, our technological revolution owes its success to the supremacy of the concept of the world as a mechanism, and man as a competitive survival machine. Everything continued to work predictably according to Newtonian and Cartesian understanding, which still informs contemporary biology, biochemistry, physics and medicine — indeed our whole worldview.

This dynamic energy field is not exactly a new idea. Vedic philosophy, three thousand years ago, postulated that matter is created from the ether that surrounds us in space, and similar theories are found in the traditions of many early civilizations.

However, the idea of the ether lost its credibility to the materialist worldview in the latter part of the nineteenth century. It was more recently revived by Einstein's protégé David Bohm, who saw the Universe as part of something vast, ineffable and essentially conscious, a creation aware of itself and its one-ness. He conceived the term 'Continuous Creation', with the idea that all of space is filled with a dynamic energy he called the etheric flux that is presumably an older term for the quantum field. We shall use the more contemporary term, quantum field, in connection with the effects and activity which take place in domains more refined than the physical. Quantum physics gives us a scientific framework for understanding the interconnectedness of all life, including our physical, emotional and mental experience.

Metaphysical science

The ancient sciences had a more holistic understanding of the world than we in the West do today. However, we do use language that differentiates between different qualities of energy. We speak of a person having 'coarse' or 'refined energy'. A product is called 'tasteful' or beautiful when it has an integral harmony and balance, but we don't use a scale of refinement.

Theosophy has a helpful schema for understanding energy.[1] It postulates a hierarchy of being or consciousness from lower to higher frequencies in terms of domains or dimensions, each separated by a 'veil' which renders the higher level inaccessible. The higher frequency is aware of the lower, but not the lower of the higher frequency.

Our normal awareness is the domain of the Third Dimension, the physical, which has its own laws. We can have glimpses of the fourth (time) and fifth (thought) dimensions, through intuition or inspiration. All subtle dimensions are present on Earth, interpenetrating the third dimension, though we are not normally conscious of them. Many other animals or humans with raised consciousness have a wider range of perception. A close relationship with a dog, cat or horse often reveals instances where the animal is aware of a non-physical 'presence' which is beyond our own awareness or which may even be a spirit presence.[2]

If our consciousness is lowered, we feel less ability to control our own lives. If all our three components of consciousness (physical, emotional and mental) are being fully used, then we can experience the full potential of being human and the gift of free will.

Quantum energy

The Nobel prize-winner David Bohm (1917-94), who was part of the renaissance of quantum physics in the post-war years, suggested a holographic model of the Universe in his book *Wholeness and the Implicate Order.* In holography, a special plate exposed to a laser beam (coherent light) produces a three-dimensional image of the subject. What is remarkable is that, if the holographic plate is smashed into a thousand pieces, each fragment will reproduce the whole picture.

According to Bohm, the vast reality that lies beyond our senses is an undivided and coherent whole, which he calls 'implicate' or 'implied'. He insists that every element in everyday material reality — explicate, manifest — contains all the information of the implicate order. He saw the boundary between the two orders like the veil (*maya*) of the ancient Vedic tradition that needs to be pulled aside to reveal the full nature of reality. This was practised by the ancient spiritual traditions through meditation and mystical trances. Their science studied the unseen as well as the manifest — it was a holistic system.

The study of fractals is also based on a similar holographic principle. They are a mathematical description of biological self-repeating cycles, a vital part of the evolutionary process. These beautiful organic-looking structures are found in fern development, in the branching patterns of trees, in blood vessels, and in the larger environment — in weather systems and coastline features.

Traditional Western science and quantum physics are worlds apart in their understanding of how the organism functions. The Newtonian model visualizes central control by the brain and nervous system, with energy dissipation according to the Second Law of Thermodynamics, and the neo-Darwinian model holds to the competitive, random nature of life systems.

Quantum physics on the other hand describes the organism's main feature as that of wholeness based on intense intercommunication of all its parts, resulting in cooperation and reciprocity, local freedom and cohesion of the whole. Mae-Wan Ho (see further below) likens the mechanistic view of matter to dead matter, while the holistic view studies the living fabric of life.[3]

The weirdness of water

Strangely, you will find almost nothing about the role of water in the organism in biology and biochemistry textbooks. That it behaves like an organism itself is one of the weirder ideas about water. Viktor Schauberger, whose theories about water have proved to be remarkably prescient, called water an organism. But, as an intuitive, he did not expand on this very clearly.

Conventionally, an organism has the ability to reproduce. However, Mae-Wan Ho, who specializes in the study of the organism, is more interested in other important indications of an organism's aliveness, such as its sensitivity to cues from the environment, and the efficiency with which it transmits dynamic energy within its bounds; both of which water demonstrates. An organism operates with long range order and coordination, as does water in certain (quantum) conditions. Water departs from Ho's qualifications for being 'alive' in that it does not have individuality or independence; however it shows a certain wholeness and coherence in the manner in which it imparts these to what we normally consider to be organisms.

The key to the quantum qualities of water, Ho suggests, is the proposition that water comes in two states — bulky, low density (super-cooled water) and high density, when the molecules are packed more closely together.[4] In its dense state, it seems to have more long-range coherence, and exhibit quantum qualities (memory, high dynamic energy, communication, self-refining, etc.). The form we know is low density water. Her research demonstrates that an organism functions by all its parts working together coherently. This coherence affects all modes of the organism, its states, phases and motion, and is governed by the essential cohesion of the water medium.[5]

Biological water

Most of our biological water is part of the inter-cellular matrix that governs metabolic functioning and chemical reactions. It is often called intercellular water, but what does it actually do? This water is in a crystalline state called the liquid crystalline continuum in which all the molecules are macroscopically aligned to form a network linking up the whole body. This continuum is diffused through the connective tissues, the extracellular matrix, and into every single cell. All the molecules, including the water, are moving coherently together as a whole, even when the body is at rest.

We are here in the realm of quantum physics, which operates under different principles from those of Newtonian physics, the main feature of the quanta being how they inter-connect all life-forms into a coherent whole. This operates from the micro to the macro level, and is very much what makes an organism tick. Conventional physics, on the other hand, focuses more on individual molecules and is less able to see the larger picture.

Mae-Wan Ho pioneered this exciting research using a polarizing microscope with special settings. With this she has developed a method of filming the behaviour of the molecular structure of organisms. Her Institute of Science in Society has produced a video called 'Quantum Jazz' which features the *Daphnia*, a fruitfly larva (it could just as well be us), in glorious technicolor, whose molecules dance around as if part of a ballet. This technology displays the changing subtle energy effects in the watery domains of a living organism. Mostly we have to study the interior of organisms when they are dead. This effect has been filmed and everyone can see it on a simple DVD. It is quite electrifying, and demonstrates, more than can a thousand words, how living organisms operate with remarkable coherence.[6] (See Plates 4, 5, 6.)

What is life?

As Ho recounts:

> To see it for the first time was a stunning, breathtaking experience, even though I have yet to lose my fascination for it, having seen it many, many times subsequently. The larva, all of one millimetre in length and perfectly formed in every minute detail, comes into focus on the colour TV monitor as though straight out of a dream.
>
> As it crawls along, it weaves its head from side to side, flashing jaw muscles in blue and orange stripes on a magenta background. The segmental muscle bands switch from

brilliant turquoise to bright vermilion, tracking waves of contraction along its body. The contracting body-wall turns from magenta to purple, through iridescent shades of green, orange and yellow. The egg yolk, trapped in the alimentary canal, shimmers a dull chartreuse as it gurgles back and forth in the commotion.

A pair of pale orange tracheal tracts run from just behind the head down the sides, terminating in yellow spiracles at the posterior extremity. With the posterior abdomen, fluorescent yellow malpighian tubules come in and out of focus like decorative ostrich feathers. And when highlighted, white nerve fibres can be seen radiating from the ventral nerve cords.

Rotating the microscope stage 90° caused nearly all the colours of the worm instantly to take on their complementary hues. It is hard to remember that these colours have physical meaning concerning the shape and arrangements of all the molecules making up the different tissues.

It was some time before we realized that we had made a new discovery. The technique depends on using the polarizing microscope unconventionally, so as to optimize the detection of small birefringences or coherently aligned anisotropies in the molecular structures of the tissue.

There is no conductor or choreographer. The organism is creating and recreating herself afresh with each passing moment, recoding and rewriting the genes in her cells in an intricate dance of life that enables the organism to survive and thrive. The dance is written as it is performed; every movement is new, as it is shaped by what has gone before. The organism never ceases to experience its environment, registering its experience for future reference ...[7]

The coordination required for [humans to achieve] simultaneous multiple tasks and for performing the most extraordinary feats both depend on a special state of being whole, the ideal description for which is 'quantum coherence'. Quantum coherence is a paradoxical state that maximizes both local freedom and global cohesion.[8]

Quantum coherence

Quantum entanglement is the term used for a high order of coherence or integrity, displayed by a shoal of fish or a flight of birds which, suddenly changing direction at breakneck speed without colliding, behave as a single organism, one part in complete harmony with every other part.

The Institute of Science in Society has pioneered research into how organisms actually work. Dr Ho explains why Nature does not recognize the Second Law of Thermodynamics. Natural systems work like wheels within wheels:

> The perfect coordination required for simultaneous multiple tasks in everyday life and in performing the most extraordinary feats both depend on a special state of being whole, best described as 'quantum coherence'. Quantum coherence is a paradoxical state of wholeness that's anything but uniform. It is infinitely diverse and multiplex, it maximizes both local freedom and global cohesion.
>
> The quantum coherent organism ... is a domain of coherent energy storage that accumulates no waste or entropy within, because it mobilizes energy most efficiently and rapidly to grow and develop and reproduce. Not only does it not accumulate entropy, but the waste or entropy exported outside is also minimized.
>
> Part of the secret for quantum coherence is that the life cycle itself contains many cycles of activities within. These cycles of different sizes are all coupled together so that activities yielding energy transfer the energy directly to activities requiring energy, losing little or nothing in the process.[9] If you look inside each small cycle that makes up the whole life cycle, you will see the same picture as the whole; and you can do this many times over until you come to the smallest cycle.[10]

The molecules are embedded in a water medium, which maintains them in a dynamic crystalline state, their electrical polarities forming a continuum that links the whole body. What is remarkable is that all the

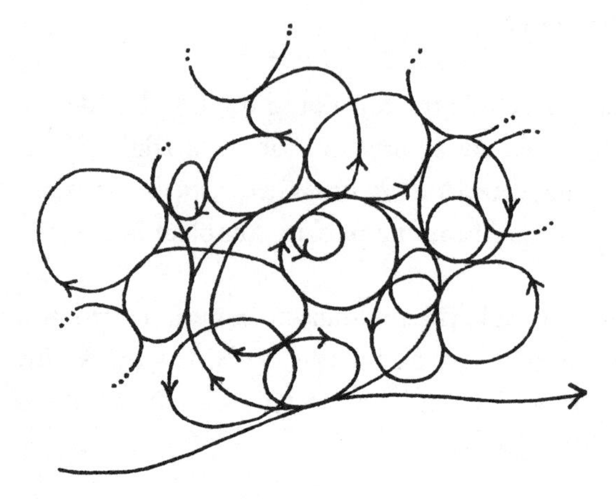

Figure 17. Organic cycles. Mae-Wan Ho's representation of how an organism's energy is continually recycling.

molecules are dancing together, and the more coherent their movement, the brighter their colours.

The high degree of coherence is dependent on the liquid crystalline nature of the water medium, which makes up about 70% of the total weight of a living organism. This allows all the molecules to intercommunicate and synchronize with each other. In Ho's words:

> I call the totality of these activities 'quantum jazz' to emphasize the immense diversity and multiplicity of players on all scales, the complexity and coherence of the performance, and most importantly, the freedom and spontaneity of it all.
>
> Quantum jazz is played out by the whole organism, in every nerve and sinew, every muscle, every single cell, molecule, atom, and elementary particle, emitting light and sound with wavelengths from nanometres to metres and kilometres; spanning a musical range of 70 octaves or more, each improvising spontaneously and freely, yet keeping in tune and in step with the whole.
>
> Quantum jazz is written as it is performed; every movement is new, shaped by what has gone before — though

> not quite. The organism never ceases to experience her environment and taking it in for future reference, modifying her liquid crystalline matrix and neural circuits, recoding and rewriting her genes.
>
> Quantum jazz is why ordinary folks can talk and think at the same time, while our breakfast is being processed to give us energy. It is why top athletes can run a mile in under four minutes, and kung fu masters can move with lightning speed and fly effortlessly through the air.

It is possible that this quality of quantum coherence could be the explanation for how the human mind can influence events (outcomes). It might also account for the accomplishments of Hindu fakirs or even for the miracles of Jesus. It would depend upon which level you are able to tune in to. Healers work on the quantum level which has different laws that govern outcomes. It is generally understood in their profession that they might lose their gift if they give in to the temptations of the ego. By doing so they would lose coherence with their client. This could also be stated as: In order to work successfully with fifth dimensional energies, it is necessary to have integrity, and a willingness to work within spiritual laws.

It has been suggested that humanity may have this gift of coherently visualizing outcomes as part of the free will package, which perhaps other animals were not 'given'. If coherence can be equated with consciousness, then the future of humanity on this Earth must be connected with the raising of consciousness of a significant number of people (a spiritual revival).

Interfacial water and water's skin

All organisms have a skin which performs a number of important functions. As the outside layer, it defines the integrity and coherence of the organism and limits its vulnerability to physical assault and infection. It is the vital heat-balancing organ for most animals. The skin is also full of tiny sensors; in some animals these represent their main 'antennae' for picking up information from the surrounding environment.

The Earth probably fulfils more completely than water Mae-Wan Ho's criteria for an organism. Its skin is the biosphere, and it is particularly poignant in these times that Earth is suffering as a result of the damage humanity has wrought. However, water's quality affects all of life. Schauberger said that when we allow water to deteriorate, all of life suffers.

Water also has a skin. The surface in contact with the atmosphere is called the meniscus whose stiffness or viscosity allows water bugs to sit on its surface without falling in. Its high surface tension allows it to be lifted up by inserting a glass rod. The meniscus is pulled up at the edge of a glass of water by cohesion, forming a concave surface, but if you spread a hydrophobic (water repelling) oily layer on the glass, it forms instead a convex boundary. Dr Gerald Pollack, at the University of Washington, Seattle, calls this skin or part of a water body in contact with another medium surface, 'interfacial water' or an 'exclusion zone' (EZ), because it seems to be able to exclude solutes that are found in the main water body.[11]

But the stream also has a skin where it contacts the streambed. The total water-body skin contains the integrity of the whole. The water molecules next to the skin usually have a concentration of active oxygen, which acts as a neutralizer of elements in the water-body that are out of balance. The water-body can retain its freshness by circulating its flow, concentrating the higher quality in the centre.

Water's skin, just like the skin of normal organisms, is sensitive to both terrestrial and cosmic energies. The meniscus has a high surface tension (ST), which is the ability of water to stick to itself. It is ST that makes water want to form a sphere — the form with the least surface area for its volume, requiring the least amount of energy to maintain itself.

In common with cell water, EZ water has complex ordered layers, which are sometimes called liquid crystals. Mae-Wan Ho discovered that cells and indeed organisms are liquid crystalline in structure.[12]

What Ho found is that this EZ water forms at every face of the water body, so that it is indeed like a skin preserving the water body's integrity. So perhaps the river's skin can absorb as well as project subtle energy from the water body into the banks and bed of the river. Could it be that water's skin acts like an antenna to receive and transmit subtle energy, as the skin does with a more 'normal' organism?

A link can be made to the restructuring of the water filaments in a stream by the action of the longitudinal vortex described by Schauberger, who said that this complex water structure enabled the water body to absorb higher energies and healing qualities, for the different layers of the water structure act like 'skins'.

An electrical field will improve water's stiffness and crystalline nature by perfecting the alignment of its molecules. Mae-Wan Ho demonstrates this strikingly on her kitchen table by putting two nearly full beakers of water almost touching; then inserting a positive electrode from a power pack into one beaker and a negative into the other. A bridge of stiff water will form, connecting the two beakers and conducting electricity. Moving the beakers several centimetres apart will still maintain this bridge.[13]

Brain consciousness versus body consciousness

The brain controls the central nervous system through the cranial nerves and spinal cord, and the peripheral nervous system. Biology traditionally identifies the brain as regulating virtually all human activity, including involuntary actions such as heart rate, respiration, and digestion. This is a mechanistic view of the human body systems which also assumes that the brain is the seat of consciousness, but brain science does not have an easy answer for how the brain is able to operate as an integrated whole. Is this not really the same question as how the organism operates as a coherent whole?

Biologists have long wondered how an organism like a bear or a human is able to respond so quickly to outside stimuli. For years it was assumed that the body's nervous system was responsible for passing messages from eyes to brain to the hands or other parts. But careful measurements show that the nervous energy paths can take large fractions of a second, too long for the instantaneous response that is usually the case.

It is a well-attested fact in the practice of aromatherapy that nutrients applied to the skin are very quickly conveyed all over the body. A herbalist friend rubbed a clove of garlic on the sole of his baby's foot, and was amazed to detect the smell of garlic on its breath half a minute later! This is possible only though the link chains of the water's magical crystalline structure.

Mae-Wan Ho gives the example of the accomplished pianist's hand-eye coordination: 'There simply isn't time enough, from one musical phrase to the next, for inputs to be sent to the brain, there to be integrated, and coordinated outputs to be sent back to the hands.'[14]

Collagen and colloid crystals

The most exciting discovery in recent years has been the strange role that water plays in biological communication. It has come to light that collagen, the connective tissues that make up the bulk of all multi-cellular animals is crucial to the integrity of the organism. These tissues are composed of a crystalline matrix of collagen proteins embedded in water which is 60–70% by weight. This, suggests Dr Ho, makes the connective tissues the ideal medium for communication. This water is specially structured in chains along the collagen fibres and has the ability of self-organizing.[15]

Tests done by Gary Fullerton at Texas University, San Antonio, suggest that water associated with collagen exhibits a high degree of quantum order, being structured in regular chains along the collagen fibres.[16] This would facilitate a process called 'jump-conduction of protons' that would enable instant communication to take place between different parts of the body — essential for perfect coordination.

Dr Ho believes this would enable water associated with collagen to become super-conductive, the ideal medium for instantaneous intercommunication for coordination of all cellular activities. She believes that this liquid crystal continuum constitutes a 'body consciousness' that may well have evolved before the nervous system, but which today works both in partnership with and also independently of the nervous system. 'This body consciousness is the basis of *sentience*, the pre-requisite for conscious experience that involves the participation of the intercommunicating whole of the energy storage domain.'

The body's energy, she points out, does not follow the conventional laws of thermodynamics, but is able through quantum coherence to be stored dynamically in a closed system, to be available at a moment's notice.

Dr Ho suggests that the acupuncture meridians of Chinese medicine may be structured water lines aligned with collagen, and that *chi* energy may be the positive bio-electric currents carried by the jump-conduction of protons through the hydrogen bonds of water molecules.

Another area where water can produce a high degree of order is with colloid crystallization. Colloids (suspensions made up of minute nanoparticles) were always thought to be homogeneous. Norio Ise in Osaka, Japan was able to create significant crystal forms in polymer solutions. The outcome of this research has been the development of a wide range of industrial applications in electronic and photon chips.

Transplants

We tend to think of the organs of the body like engine parts that can be replaced when they wear out. Indeed, many lives have been saved by organ transplants from one body (usually at death) to another. Besides the vital organs, bone marrow and blood are also transplanted or transfused.

But there is a downside to this practice. The human body is not just the sum of its parts. It is an organism which acts as a complete unity. The heart and blood are as individual as a person's brain, and unforeseen consequences may result from a blood transfusion.

Viktor Schauberger regarded blood as an organ. There is anecdotal evidence of personality change in the person who has had his blood supply replaced by another's. Blood is similar to water in its ability to carry information. Perhaps it also has a memory. When you think about it in this way, it is not surprising that some religions regard blood transfusions as unethical.

12. Spirals, the Vortex and the Etheric

In the beginning was the vortex.
Democritas (460–370 BC)

We live on a planet that is hurtling through space at a breathtaking speed while spinning on its axis, and we are also subject to gravitational and magnetic forces. The fluids of Earth are particularly affected by the Earth's spinning. As water is the most important constituent of life, the way it moves, its rhythms and pulsations are at the heart of all life's processes.

Self-organizing systems

The linear or mechanistic system of Newtonian physics has dominated the natural sciences for three centuries. Its shortcoming is that it studies only predictable phenomena, which is why we still know so little about water. Newton did not recognize that living systems behave in what seem like a random, unpredictable way

The idea that organisms are self-organizing was first suggested by the research of Boris Belousov, a Soviet chemist. In 1959 his crucial experiment was to prepare a solution of some thirty chemical substances, ranging between a colourless liquid and one with brilliant colours which, to his surprise, organized itself into regular patterns of spirals and vortices.

These self-organizing oscillations in water demonstrated that its chemical reactions are unpredictable. Belousov's research was not taken seriously, and eleven years elapsed before a young chemist called Anatoly Zhabotinsky repeated the older man's experiment, which then became known as the Belousov-Zhabotinsky reaction. This was a challenge to the law of entropy, showing that, from a disordered state, spiral-forms emerge creating stable, oscillating patterns — order spontaneously emerging from chaos.

Ilya Prigogine (1917–2003), winner of the Nobel Prize for Chemistry in 1977, believed that organisms, though stable, are always striving towards greater equilibrium.[1] He saw organisms in a state of constant change, responding to other organisms in order to achieve harmonious balance. He acknowledged that his ideas were inspired by the Chinese tradition of fluctuation between *yin* and *yang* tending towards harmony of the Tao. Nature's imperative is to follow patterns that lead to equilibrium. This principle is also at the heart of Schauberger's work with polarities.

An organism is not a fixed entity. As Mae-Wan Ho shows in her Quantum Jazz (see Plates 4, 5, 6), organisms are in a state of constant flux, with a tendency to cooperate with other organisms in order to create an ordered reality and higher symmetry.

Order out of chaos

The principle has been recognized by the philosophers of all the great civilizations — that change can come only from the failure of the status quo. Put another way: order must be preceded by chaos; the two are inseparable partners.[2] In the Chinese *Tao Te Ching,* the hexagram for 'crisis' is the same as for 'opportunity'.

Water stage-manages life, but water itself is a disorganized medium. It seems to thrive on paradoxes. The asymmetry of its molecule is responsible for many of the strange anomalies that make it fit for life. This imperfect symmetry holds the secret of matter's very existence, for instability and imperfect symmetry are the guiding law of the Universe since its inception.

Evolutionary initiatives seem to have a tentativeness about them (trying it on?) which would translate as instability. It is the restlessness of water which gives living systems the ability to become ever more complex and to strive towards the perfection they can never reach. As the driver of evolution, water is a model for our own striving, allowing us to evolve through our own mistakes.

It is the changeability and restlessness of water that drives evolution, for it allows life, by stops and starts, to become ever more complex.

Organisms are self-organizing closed energy systems, but they also allow energy to pass through them.

We love to disparage meteorologists for getting their forecasts wrong, but weather systems are complex, non-linear systems. The watery domain, gaseous or liquid, is essentially unpredictable. Organisms that are based on water are non-linear systems, in constant flux. They communicate with other organisms, as the over-riding imperative in Nature is to bring about balance and higher order.

Turbulence and chaos

One of the main features of water is its restlessness. It is constantly opening up to new impressions, and consolidating them through turbulence and restructuring.

We normally think of chaos being like the state of a room after a wild party, or a madman's mind. Its converse, order, which is to do with predictability, straight lines and mechanical laws (like the swing of a pendulum), is abhorred by Nature. It is stifling to creativity. Chaos theory shows us that this apparent 'chaotic' disorganization is a state open to the emergence of anything new; it is the way Nature creates new order. It is the driver of evolution.

The key to this new order is the fractal, the self-repeating structure that conveys to the smallest scale of life the patterns of the macro Universe, through the action of the quantum field working through water.

Computers and abstract mathematics, working with the broader view of quantum physics, have given us a deeper understanding of chaotic systems. Contrary to what conventional physics might expect, this apparently random chaos is actually a precondition of higher order and intelligence.

You can extrapolate this idea into life in general. Any artist will tell you that inspiration does not arise out of the predictable, but from the unexpected, the disturbing situation. A routine, predictable life lacks the creative spark that enables one to discover the best opportunities for growth.

Recognizing this can transform the distress of a relationship break-up or loss of one's job into an opportunity for creative change. The ancient

Chinese saying 'Breakdown brings opportunity' is really quite apposite. This is what water teaches us, as turbulence allows water to transform towards increased complexity and higher order. 'Going with the flow' is about having faith that the life process has behind it, a wisdom about balance and growth. The great evolutionary leaps came out of chaotic geophysical events. You might even say that human evolution is helped by water's restlessness, which encourages us to learn from our mistakes and seek to become more whole (see p. 279).

Are we able consciously to allow Nature to break down our collapsing and unsustainable human organizations with the belief that new, positive structures will evolve?[3]

The vortex

The Earth's rotation causes all fluids and gases to move in spirals. Water's spiralling is familiar because it is the more visible (Plate 16). But you can see it with smoke and sometimes with mists. Spirals are a basic form of motion in Nature, but Schauberger in the 1930s recognized the vortex as the principal creative movement system in the Universe. It is at the core of his eco-technology and the condition required to produce the vortex is turbulence, or chaotic motion. Turbulence increases with a small rise in the temperature of the water, which will disturb a fish and make it more likely to go for the fisherman's bait. Turbulence is the precursor to chaotic restructuring of the water.

The vortex we see as the bathwater goes down the plughole is comparable to the structures of cyclones, of DNA, magnetic force fields and galaxies. It is a concentration of spiralling and dynamically increasing energy. A healer's subtle energy projects from his hands as a vortex of concentrated energy. A tornado's energy increases with the narrowing of the cone down to the point, where the tornado's destruction is greatest.

It was Lord Kelvin (1824–1907) who first proposed that atomic motion is vortical. This was picked up by James Clerk Maxwell (1831–79) and by Sir Joseph J. Thompson (1856-1940) who discovered the electron.[4]

The idea fell into disfavour when the concept of the ether was discredited (the theory that atoms required a substance to move in).

The concept of wave motion took over, and the atom was demoted from its position as the smallest unit of matter.

It took the emergence of String Theory in the 1970s and 1980s to bring back the idea of atomic movement in spirals. It describes how dynamic energy can create what looks like static matter by conceiving a ball of string composed of countless short threads each forming spirals intertwined with left and right hand spins, echoing the electromagnetic polarities found throughout the Universe. The vortex was rediscovered by quantum theory.

DNA is probably the most famous spiral in Nature. It is also the battleground between Neo-Darwinists who explain evolution through random mistakes in replication of DNA and those who interpret evolution as an imperative process of Nature's intelligence.

James Gleick, who popularized the ideas of chaos theory in the early 1980s, pointed out that life is created from chaos. Fluids are the ideal media of chaos because of their unpredictable and turbulent behaviour. Water turbulence creates vortices and spirals in an energetic stream, with the flow direction constantly changing from left to right.

From the tornado to the way a plant grows, the vortex is Nature's mechanism for increasing the quality of energy, raising it from a lower to a higher level. It is a powerful tool for evolution, and perhaps water's most constructive role is to use the vortex for this purpose.

Ilya Prigogine saw that disorder creates simultaneously stable and unstable oscillating systems that are spiralform in structure leading to order. This oscillation enables a dynamic homoeostasis (stability) to exist, and to increase complexity and energy levels. This mirrors Schauberger's practical research in how a stream raises energy levels which he applied to his eco-technological applications. Prigogine sees the vortex as energy dissipating, in line with Schauberger's view.

The vortex is a window between different qualities or levels of energy. Black holes can be thought of as vortices, gateways linking different parts of our universe or even different universes. The vortex and spiral have become hallmarks for Viktor Schauberger, as for him they were the key to all creative movement. As we shall demonstrate later, the vortex is most clearly seen with water, which it uses as vortical motion to purify and energize itself, introducing finer etheric energies to wipe clean the bad energies of the water's previous memory of misuse.

One could use the image of a musty room that feels stale and unwelcoming. Once sunlight and fresh air are allowed to penetrate, the unpleasant atmosphere is quickly transformed. It is a natural principle that the more refined energy always prevails over the coarser. As Viktor Schauberger demonstrated, Nature's evolutionary 'purpose' is continually to refine and to create greater complexity and diversity, the vortex being the key process in this endeavour.

Eggs and vortices

Until comparatively recent times, scientists and philosophers recognized the creative energy of Nature as sacred. They saw the way in which Nature's patterns and its complex interdependences were often expressed in very specific shapes that could be described in geometric terms and ratios, as proof of the hand of God at work. So they called these correspondences sacred numbers and sacred geometry.

In art and architecture, the search was for true balance, perfect proportion, a shape that is aesthetically pleasing, The square is too mechanical, a long rectangle too awkward. The shape that 'seems' to be just right is the square rectangle with the proportions of 5:8. It was called the 'Golden Mean', usually described by the Greek letter phi (φ). This turns out to be the magical proportion favoured by Nature in her designs. It is also the key to the shape of the egg.[5]

Viktor Schauberger was well aware that Nature uses the egg shape for creating and maintaining the dynamic energy of biological water. In his *Fertile Earth,* he wrote: 'Every force ... unfolds itself and springs forth from the original form of life, the egg.'

He understood also the importance of the electromagnetic polarity being activated in water. For example, in generating vortices in egg-shaped vessels as a way of developing powerful energies in applications for heating and cooling and for high potency drinking water he placed the pointy end up. This produces a negative *yin* subtle energy like the in-drawing of breath. When he used eggs for potentizing natural fertilizers he placed the pointy end down, producing a *yang,* out-breathing subtle energy. He did not, as far as we know, experiment with two eggs used together, but he demonstrated that an egg with the pointy end down generates dynamic energy, while with the point up gathers in, condenses and concentrates energy.[6]

Ralf Roessner, a German anthroposophical scientist, has produced a model which ingeniously separates the *yin* and *yang* parts of the process, illustrating clearly what takes place. He uses two glass eggs, joined at their pointy ends by a short collar, to produce high quality drinking water (see Plate 7). It is an ingenious system: the upper egg produces a fine vortex and the water accelerates and chaoticizes as it passes through the narrow neck. When it falls into the lower egg as turbulent and chaotic myriads of tiny electro-magnetically polarized cyclones, liquid crystalline chains form and a natural clustering takes place which rebuilds the internal structure of the water at a higher dynamic energy level. The surfaces of the many layers in this laminar structure act like skins or antennae to absorb higher energy.

This dual system elegantly combines the energy dissipating function of the vortex with the energy enhancing mode of chaos in the lower egg. The natural egg shape in each case encourages the water body to circulate efficiently as a coherent whole. The natural coherence of water is amplified by the egg shape, helping it break through into the quantum domain, from which it can absorb higher quality subtle energies. Quantum coherence is much more dynamic than normal physical coherence.

Roessner spent ten years developing this double-egg system, with beautiful hand-blown eggs, created individually by skilled glass craftsmen, to produce quantum water for domestic use.[7] He points out that through this very personal, hands-on way of energizing water, we can communicate with the water being energized. As we are essentially quantum water, our own biological water is involved with the process of spinning the water in the eggs.[8]

Flowforms

We haven't talked much about rhythms with water. In the flowing of a healthy stream there is a natural rhythm of movement from side to side, which keeps the electromagnetic charge of the stream in balance. If you wish to create this rhythm in an artificial stream, it is necessary to understand the mathematics of flow in relation to form.

John Wilkes, as a sculpture student, was introduced to projective geometry by the anthroposophist George Adams, a colleague of

Theodor Schwenk, whose pioneering research we discuss in the next chapter. He believed that the quality of water may be dependent on the rhythms of moving water on the surface over which it flows. From this came the idea of the flowform, which consists of a series of heart-shaped bowls that introduce rhythmic movement into the water in figures of eight. It is usually built as a cascade of these bowls, in a wide variety of shapes and in different materials (often pre-formed concrete). (See Plate 9.)

Over one thousand flowform projects have been set up in over thirty countries. Their ability to invigorate the environment has made them popular in town centres, large businesses, parks and gardens. Their most practical use is in conjunction with water treatment plants, where they play a crucial role at the end of the purification process. They are also used in biological treatment systems, fish farms and reedbed sewage systems.

The ground of all being

Goethe called water 'the ground of all being'. 'Being' is more than manifest life, it is also the ground of Bohm's 'implicate order'. By the Law of Polarity, all systems have a dual aspect (see p. 282). Thus it is reasonable to suppose that cosmic principles and forms are conveyed to organic life as a partnership between water (*yin*) and the quantum field or the etheric (*yang*).

Both are media for the transmission of subtle energies. Whereas the etheric may have to do more with the mental level, water is perhaps the medium for communicating formative information. It seems that they work together in harmonious balance.

I suspect that the quantum field may be related to electromagnetic energies as they are generally understood, but on a higher dimensional level and therefore presumably subject to different laws; for example, while normal electromagnetic waves are interrupted by lead shielding or a Faraday cage, the etheric may not be. These are still early days in the understanding of the etheric field, an area of study that might be facilitated by the concept of a partnership with the water domain.

13. Water's Cosmic Role

Like all living things, water is a self-organizing system: a drop of water is a universe unto itself, containing substances and living creatures, generating and responding to vibrations, perpetuating the same patterns and sequences that we find on a larger scale in our own bodies, in the Earth, indeed throughout the entire universe. [This drop of] water is the smallest, yet also the most all-encompassing and immanent of universes.

Paolo Consigli

The cosmic connection

From earliest times, Man has believed that the heavens influence life on Earth. In earlier cultures, like the Babylonian, there were deities to be placated. Gradually, however, the study of cosmology developed; science and religion were inseparable, being part of the skills of the priesthood. The famous maxim: 'As above, so below', attributed to Hermes Trismegistus, the mysterious founder of alchemy, refers to the theory of correspondences: 'That which is in the lesser world (the microcosm) reflects that of the greater world or universe (the macrocosm).'

The 'harmony of the spheres' is the poetic phrase that recognizes a natural order in the motion of the planets in the Solar System and of stars in their constellations. All life is motion. Natural movement comes not in straight lines but as spirals, the shape taken by fluid energy evolving from chaos (Plate 17). Patterns and rhythms are the heartbeat of the Universe. They are to do with order, with design and structure. Nothing can come into being without a design or template. How are these templates or patterns for growth and harmony conveyed to the physical Earth environment?

Ancient cultures recognized mathematics and geometry as the tools to understand patterns in Nature and in the Universe. It was from such relationships that 'the Pythagorean canon of proportions' was created. The basics of musical harmony depend on intervals created by these divine proportions. Canons of architecture, of painting and of musical harmony were taught in the medieval mystery schools, and partly revived in the Renaissance. Understanding the mystery of these forms is a whole science in itself (sacred geometry)!

Perhaps the most important role that water plays is that of transducer between the cosmic and earthly realms. There is a vast circulation system in the crust of the Earth (nearly one third of the Earth's fresh water is very deep underground). Rain charged with the cosmic energy penetrates the Earth's surface and joins this circulation, to be charged with minerals and with the mother energy of the interior.

Cosmic energy is also absorbed into the water system by plants and directly by the land. This requires antennae, such as bud forms, seeds, egg shapes and tree leaves (esp. pine needles), or certain points in the landscape (for instance, powerful hills or mountains).

As well as energy transduction and supply of nutrition, water carries information that is the very basis of the web of life. The quality and amount of information carried depends on the structure of the water.

Resonance

All matter, though it may look solid and stationary, is based on subatomic particles which are always in motion. The velocity of this motion determines its vibrational rate; this and the type and size of the object contribute to its vibrational or resonant frequency.

Life-forms in Nature respond to each other by means of resonance — 'Gaia's glue'. It is the language of communication and response. Resonance is what holds Nature together; it is the law of attraction, bringing the lichen to the rock, the orchid to the tree, the butterfly to the buddleia, the bee to the blossom. Resonance is similar to Mae-Wan Ho's concept of coherence (see Chapter 11).

The resonance and vibrational rate of water is at the heart of natural medicine (esp. homeopathy), biocommunication and husbandry. This works through sap and blood and controls how organisms grow. The quality of the water affects both recharging and cleansing.

Sound is probably the most ancient form of resonance in human experience. Jericho was reputedly destroyed by destructive sound resonances. A piano tuner uses resonance. Music itself is more than a paradigm of the forces behind Nature. For millennia people have sung and played music to their crops, their streams, their loves, their children and their animals.

When we experience a sense of the sublime on listening to a musical composition of great integrity, we come into a state of resonance with the particular dynamic energy pattern of that composition. The same can happen with the thrill of watching the aurora borealis or similar natural phenomena — one's whole body resonates with that particular energy (Plate 18).

Of course you can resonate with degraded energy too, particularly if one's own dynamic energy is degraded.

Cosmic influences: Rudolf Steiner and Theodor Schwenk

One of the best philosophical sources for understanding the greater cosmic system or 'consciousness' is the body of knowledge called Anthroposophy, established by Rudolf Steiner (1861–1925). Steiner was an inspired polymath, whose principal concern was the relationship between Man and the Cosmos, about which he constantly sought spiritual insights. He made no secret of being over-lighted by spiritual beings. He was influenced by Goethe who believed the Earth to be an intelligent organism, as did Johannes Kepler, the great astronomer.

Also a practical reformer, Steiner was a pioneer of holistic education for children, of a radical type of homeopathy and a new style of Christian worship. There was also biodynamic farming (see pp. 270), a new understanding of horticulture, now widely practised in Australia and New Zealand and in parts of North America and Europe.

What concerns us in this study is his concern with planetary influences on organic life and agriculture, in sacred geometry and correspondences between the Cosmos and living organisms. Steiner believed that our organs have different roles as receptors of cosmic or planetary energy.

We have established that water quite probably exists throughout the Universe. Its role, says Steiner, is to transfer information necessary for any particular part of the web of life for its growth, development and connectedness to the whole, conveying to Earth's biosphere patterns of the Cosmos, its rhythms and resonances.

Before the mechanization of Nature imposed by the dominant Western scientific model from the late seventeenth century, peoples all over the world in traditional areas paced their lives upon cosmic rhythms. These influenced sowing and planting, irrigation, fertilizing, harvesting of crops, and the siting of buildings. The word 'rhythm' comes from the Greek word for 'flow' and suggests the need 'to go with the flow'

Many highly qualified scientists were influenced by Steiner's vision and made original contributions in their fields: for instance, in geology, holistic biology, botany, anthropology, hydrology, bioresonance and sacred geometry.

While Steiner was very interested in the rhythms of the Cosmos, Theodor Schwenk (1910–1986), who was inspired by both Steiner and by Goethe, studied how these are mirrored in water. Schwenk founded the Institut für Strömungswissenschaften (Institute for Flow Sciences) in Herrischried, southern Germany. His research on the flow of water and gases is pioneering. He also developed a method of analysing the quality of water called the water drop technique, and he had inspired ideas about water, many of which are recorded in his seminal book *Sensitive Chaos.*

Figure 18 (right). Water drop samples. These 'drop' pictures show the structure of water samples. The first is of living spring water with its structure complete; the second taken downstream after the inflow of domestic sewage and industrial effluents, with a trace of rudimentary development; a third taken still further downstream shows how the stream has, through its naturally spiralling movement, rebuilt the water's structure. (Institut für Strömungswissenschaften)

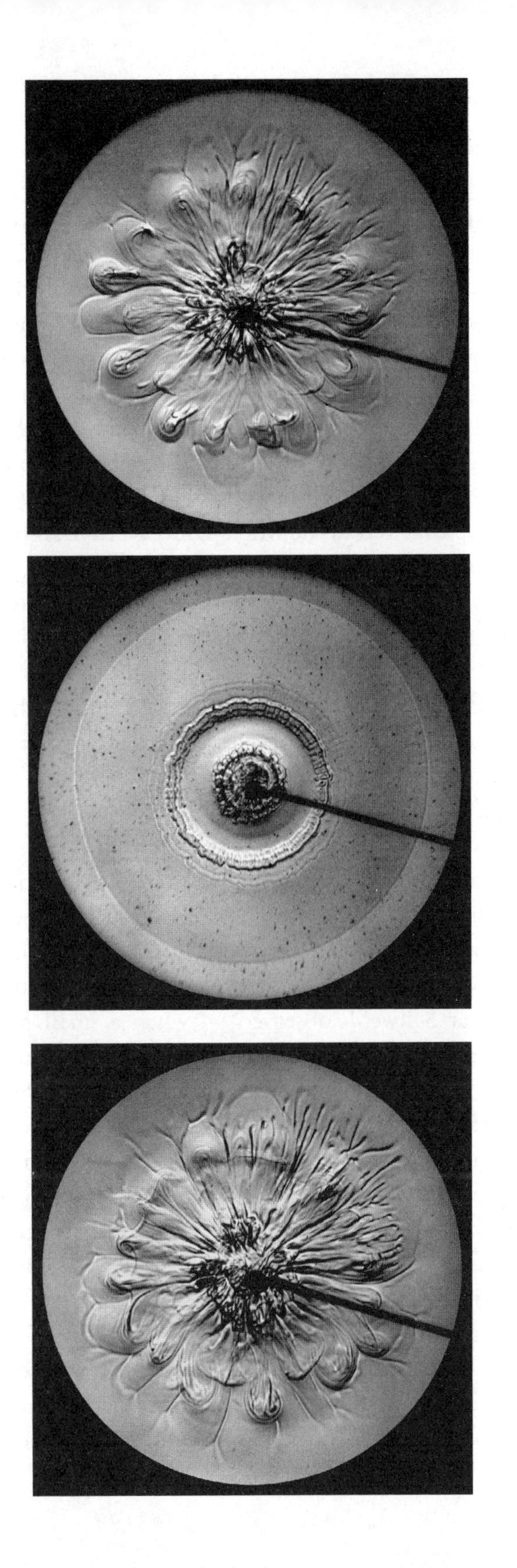

The role of water

What we are proposing is that the etheric and water are the conveyors of the cosmic intelligence, which we call 'consciousness', that has put together the intricate interconnections of the web of life, but which has required water to be self-regulating and self-evolving from the instincts of wholeness perhaps pre-programmed before the beginning of time.

Remarking that the wisdom we find working in the smallest organism or tiny organ is the same wisdom that underlies the whole environment of life, which comes from the regulating element of water, Schwenk calls water Nature's central organ, its 'heart' — the pulsing, oscillating drop that allows the whole Universe to pass through it. Just as the human heart mediates between all the other organs, so water mediates and balances the energy transfer between organisms. 99% of all chemical and other changes depend on water. As the heart balances the upper and lower functions in the human, so water balances the Cosmic and Earth energies in the Earth organism, which he also calls the spiritual and material manifestations, balanced by its rhythmical motion: 'It acts as the small eccentricity of Nature, and through this, results in life.'

The vortex is a structure that is complete in itself, with its own rhythms and movement, very similar to that of the motion of planets around the Sun in the solar system. It follows Kepler's Second Law of Planetary Movement, in that the speed of rotation is higher near the centre of the vortex than it is on the outside. There is another cosmic connection that Schwenk observed — when a small piece of wood shaped like a pencil stub is immersed in the vortex it will always point in the same direction, as the axes of planets do, circling the Sun.

Schwenk's book *Sensitive Chaos* — a title derived from the eighteenth century German mystic and philosopher, Novalis — is one of the most original in the study of water behaviour. It offers an inspiring collection of photographs and drawings illustrating how our organs and bone structure often mimic water flow patterns, giving support to the idea that water carries the template for every form of life to develop as its blueprint demands.

Steiner's vision of the pivotal place that water plays in the development of life shares much with Schauberger's. Steiner died before Schauberger really got into his stride, and there is no evidence that they met.

I do think that Schauberger's ideas must have influenced Schwenk, although the Schauberger books did not appear until after his own death. However, he probably read Viktor's numerous articles in magazines like *Implosion*.

Three characteristics of water

Schwenk recognizes three characteristics of water. The first two are easy to observe, and generally acknowledged. Firstly, there is water's vital role in all metabolic processes in the Earth, in the great aerial ocean and in every living creature. Then it is clear that water is inseparable from the rhythmic processes that take place in space and time.

As Schwenk writes:

> When water stops dancing and flowing, it stagnates, loses its life, as though paralyzed. The flowing and rhythmic patterns of water are central to Nature's essence. Thus we find the audible rhythms of brooks and of the ocean mimic the patterns of waves and meandering watercourses, and the fibrous structures of the brain mimic water flow ...
> ... Every twig dangling in a river causes a train of vortices in the rhythmic sequence; about every surface of contact between two streams there is a rhythmic play of waves and vortices. (*Sensitive Chaos*, p. 83.)

The last characteristic is that water is the antenna that picks up influences far beyond the Earth. This can be intuited only by very close observation of the inner boundary surfaces that permeate every body of moving water and which arise from the interplay of constantly changing laminar structures. They act like membranes with the sensitivity of sense organs, which made Novalis designate water as 'sensitive chaos'.

Water's sensitivity

The skin of any organism is its contact with the outside world, its antenna, its sensitivity.

Dividing surfaces or boundary layers appear not only between water and air, but also within the water mass. The essential structure of water is laminar. Moving water consists of layers flowing past each other at different speeds, causing deflection and rolling-up, which create vortices (which mark the boundary between inner and outer space). A flowing movement increases the complexity of these laminar structures or membranes, allowing them to expand, contract or make rhythmical waves.

Each of its layers has a boundary surface that responds to outside influences, even those deep within the water body. It may be that different layers pick up different vibrational levels, like tuning a radio, so that bringing together all these vibrations would form a great cosmic orchestral symphony.

These surfaces become waved, curling over and forming a circling vortex. Where streams flow past each other at different speeds there are surfaces of contact. As long as water is moving it absorbs information. When it stops moving it ceases to absorb information — like turning off a tape recorder.

Water's sensitivity is as great as that of the human ear. It is in its nature for water to be sensitive to the slightest stimuli. A gentle breeze immediately creases the surface of water into the smallest of capillary waves. A small obstruction creates a whole system of tension waves.

Then there is the influence of the Moon's rhythms on human hormonal activity. Less well known is that the Moon's rhythms affect the movement of sap in trees which changes according to the phase of the Moon and its position in the zodiac.

Timber felled at new Moon in the winter is the most durable. In parts of South America still, trunks of valuable hardwoods are stamped indicating the phase of the Moon when they were felled, as this affects their commercial value.

Even within the Earth, water rises and falls with the changes of the Moon. When digging a well, at some times one can dig shallow to reach water, at other times one has to dig deeply. Rudolf Steiner recommended planting seeds and plants at specific phases of the Moon, in order to maximize their potential for healthy growth.

Schwenk sees the vortex as following Kepler's Second Law of Planetary Movement, a system depicting a miniature of the great starry Universe:

> The vortex is a moving part within a moving whole. It has its own rhythms, forms its own inner surfaces and is connected to distant cosmic surroundings.

He says that water is not itself alive. Life manifests itself in many ways, through growth, reproduction, digestion of food and excreting of wastes, and other metabolic functions. Water has none of these, yet without water there can be no life. These attributes of life depend on water. Water does not grow because it is life itself — the universal element.

There is a constant exchange between elements of air and water. Water is the regulator of climate, meteorological processes and their rhythms.

The key to water's role in the human body is that water's specific heat of 98.4°F, exactly matches normal blood temperature. The role of water in the human body is to bathe, cleanse and nourish every cell. Our brain floats in water, which facilitates thinking. The composition of sea water and blood are almost identical, except that sea water contains magnesium and blood iron. Healthy drinking water is balanced between acidity and alkalinity.

Schwenk produced remarkable photographs of trains of vortices made by pulling a rod through a glycerine-topped body of water. The fractal model of reproduction can be seen through the repetition of smaller vortices in these trains. They are also found in Nature in the generation of smaller cyclones from parent systems.

Water as mediator between Cosmos and Earth

Schwenk studied how cosmic rhythms are mirrored in moving water, in the trickling stream, the rolling river, the rhythmical ebb and flow of the waves, in the foam of the breakers, and concluded:

> Water becomes an image of the stream of time itself, permeated with the rhythms of the starry world. All the creatures of the Earth live in this stream of time; it flows within them and, as long as it flows, it sustains them in the stream of life.[2]

All living organisms show, in their form, that they have been through a watery phase of development. Some solidify only slightly. Others leave the world of water and become dependent on earth. Does the impressionable nature of water make it sensitive to living, formative forces and creative form and expression? Perhaps water is the vehicle that allows cosmic forces to penetrate the material world in order to create living organisms. (See Figure 20.)

Theodor Schwenk tells the remarkable story of the smelt, which is distantly related to the salmon. Once a year at spawning time in May, they approach the coast of California. Patiently, they wait near the shore until the tide reaches its highest on the third day after Full Moon. Then, choosing the, highest wave, they allow themselves to be carried up onto the beach. There the females lay their eggs, the males fertilize them, and with the next, last wave, they return again to the open sea.

The high tide does not reach this level for another fourteen days, when the spawn of the smelt hatches a few minutes before being washed out to sea, not to return until years later, for a moment on the third day after the May full Moon.

Amazingly, then, the world of moving water, which is the Earth's antenna, absorbs the changing combinations of vibrations of the Sun, Moon and planets and passes on this information to the fish of the sea and the creatures of the land.

Schwenk remarks that the more we learned about the physical nature of water, the more we were able to control it in great dams, transport it in huge pipes to drive turbines. It seemed to make sense to drain wetlands and straighten the rivers. The new technology gave us a sense of power, but at the cost of ceasing to understand the true nature of water. He comments that when we are concerned only with what is profitable we do not see the connection between things; we see the individual as the important unit. The essential connections and interdependences of all activities in life are missed.

We can see these connections abundantly in Nature, but we have stopped observing Nature. The artist still observes and comments; we need to encourage our creative inner artist. The market economy glorifies individual profit; it is the enemy of Nature. Our society today values the scientist more highly than the artist.

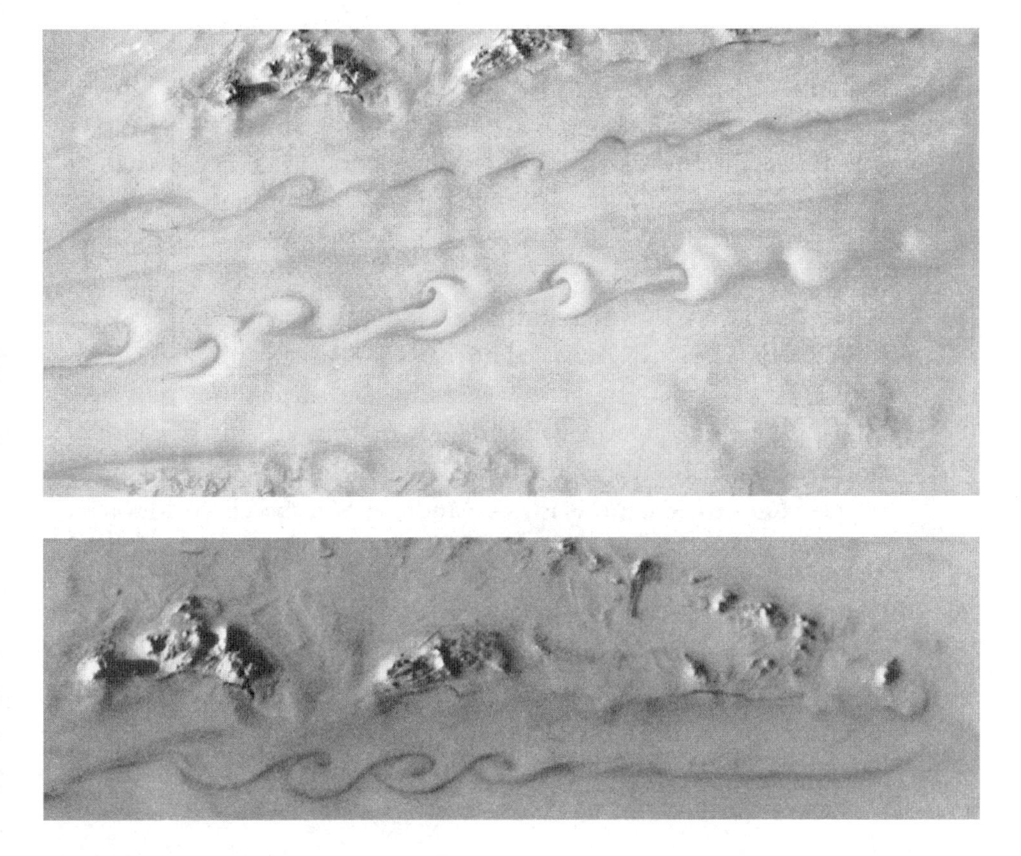

Figure 19. Beach vortices. Bell-shaped forms, which travel along with the current, arise in water flowing round a submerged stone. They are here made visible by the water stirring up the fine clay. Top (a): Bell shapes form after passing a submerged stone. Bottom (b): A train of vortices develop after an obstacle.

Templates for organic growth and evolution

Nature is a whole system comprised of two opposing qualities in balance. The Sun charges the positive (*yang*) energy requirements of life, while Earth herself provides the balancing negative (*yin*) energies. Without this interplay, there would be no water, plants, nor chemical compounds. You may remember we described the metabolism of the tree as an example of this interchange (see Chapter 8).

Pioneering research conducted by Lawrence Edwards, an anthroposophical mathematician, indicated that there are universal laws, as yet not fully understood, which guide an organism's growth into predetermined patterns. He devoted his entire life to examine in minute detail the influence of celestial motion on the growth patterns of plants, and on the form of embryos, the heart and other organs. Detailed measurement and analysis, comparing these with ideal geometric forms, led Edwards to suggest that living forms are affected by patterns in time as well as space, with all of Nature passing through vast cycles of change.

He found that many tree buds, in their dormant winter stage, are far from asleep. They are influenced by the Moon, usually in combination with another planet, to open very slightly, and then close again — a kind of 'breathing' in roughly fortnightly rhythms (that is, the Moon's full and new phases). With the oak it is the 0° and 90° angular aspects of the Moon to Mars; the ash — Moon to Sun; beech — Moon to Saturn; birch — Moon to Venus; sycamore — Moon to Jupiter, a kind of astrology of Nature.[3]

Almost all rhythms, from Moon rhythms reflected in the hydrosphere and planetary rhythms known to plant growth, right down to the many physiological rhythms found in every living organism, are based on the pulsation of water. These cosmic influences affect all living things, including us, in ways that we do not recognize.

Planting by the Moon

Planting seeds at the time of the Full Moon has been practised since time immemorial, but the science behind the belief of resulting healthier plants has been tested only in modern times. The most famous innovator was Maria Thun of Darmstadt, Germany, whose research was supported by biodynamic farmers. In the 1950s she planted seeds according to varying phases of the Moon, confirming that potatoes planted at Full Moon always did best, and those planted at New Moon invariably did poorly.

Her next discovery was revolutionary. It was that potatoes planted when the Moon was in the constellations of Taurus, Virgo or Capricorn (the earth signs), were more prolific than if they were planted in other constellations.

Dr Frank Brown of Northwestern University conducted meticulous tests over a ten year period. He found that plants absorbed more water at Full Moon, and least at New Moon. Rudolf Steiner established a relationship between the elements of earth, air, fire and water, which correspond to different parts of the plant. Earth corresponds to root, water to leaf growth, fire to seed production and air to flowers; so the indication is that root crops should be planted when the Moon is in an earth sign.

Another theory is that a waxing Moon stimulates leaf growth, while a waning Moon encourages root growth. Biodynamic methods are based on more complex astronomical positions of the Moon, taking into account the low and high points in the Moon's orbit, ascending and descending, as well as eclipses.

There's certainly something in these ideas, and as there is now a great following for biodynamic methods, one can expect interest in planting by the Moon to increase.

A living Earth

We have been repeatedly asking the question: 'What is living water?' Truly, we shall not be able to understand the meaning, the real significance, of water until we understand the wider context, that of the living Earth and the living Cosmos.

I believe that many people have a race memory of being part of Nature, even though for many centuries this has been lost to modern culture. We may be able to appreciate the idea that Nature is evolving towards greater complexity and biodiversity, but it is much harder to comprehend how Earth herself might be going through her own evolution as part of a rising consciousness in the greater Cosmos.

The concept of a living Earth has been around for a long time. Plutarch, Leonardo da Vinci, Goethe — all espoused this belief. In modern times Kepler, Steiner, Schauberger, Lovelock, Bohm and others have shown that this is more than mere theory, but can be apprehended through human experience.

An Earth process with which we are daily familiar is the weather, involving a constant exchange of water between of the vast surface of the oceans and the different layers of the atmospheric ocean. Meteorologists often find themselves describing such processes as if they were something alive. August Schmass writes of 'biological concepts in meteorology', and of an 'orchestral score' with 'entrances' in the changing seasons.[4]

Paul Raethjen writes about cyclones that 'behave like a living creature':

> Cyclones have a metabolic process without which they could not exist; they constantly draw new masses of air into their vortices and excrete other masses in their outward-spirallings ... They have a typical life history with characteristic beginning, developing and ageing phases. They reproduce themselves, not a wave-like spreading out in space, but like a living creature, in the sense that a young 'frontal cyclone' is born out of the womb of an adult 'central cyclone'.[5]

Information is knowing that water is H_2O; knowledge is being able to make it rain.

Inca saying

14. Water as a Communication Channel

Water plays the role of the untiring carrier of light, energy and heat. First and foremost, it is the carrier of all the substances that create and sustain life.
Viktor Schauberger, *Fertile Earth*

Dowsing

Dowsing (water divining) has always had a connection with water, especially spiralling water. Traditionally dowsers are skilled in finding and discovering the locality, quality and reliability of underground water supplies or sometimes minerals. This skill is much depreciated today, as dowsing does not easily pass the conventional tests of repeatability under laboratory conditions. Just like homeopathy, it is dismissed by most scientists as pseudo-science, mainly, I suspect because their template does not cover subtle energies and how they work.[1]

The august British Society of Dowsers (BSD), founded in 1933, seems to attract very practical people. Its list of past presidents includes doctors and high-ranking army engineers. Dowsing has broadened out into many fields. The BSD has a number of special interest groups: for instance, archaeology, water divining, earth energies and health. Other themes are also explored, like map dowsing, lost objects, missing persons and kinesiology.

Dowsers use a variety of instruments. Most common, for field work, are the parallel rods, which might be just a pair of coat hangers bent in a right angle, or something a bit more sophisticated. Some still use a forked hazel twig. For medical dowsing, and most indoor work, a pendulum is mostly used. The experienced water dowser is able to tell, not only the presence of water, but its amount, depth, flow, salinity, pH, and other qualities, just by asking questions internally.

In 1995, Hans Dieter Benz, a physicist at the University of Munich, published research showing the efficacy of dowsing in arid regions of the world.[2] The German government supported his project to drill over 2,000 sites in several Asian and African countries. The results were impressive — for example, in Sri Lanka, a 96% success rate of finding drinkable water.

The occupational hazard of any form of dowsing is the possibility of a self-fulfilling outcome. One of the basic premises in any kind of research work is that the attitude or mind-set of the researcher can influence the result of the experiment, a truism, unfortunately not widely recognized in conventional scientific protocols. An essential part of the training of any dowsing hopeful is how to reduce this risk.

It seems that many dowsers have a facility for one area over another. For example, I am drawn to map dowsing and site dating, but am not much good at water divining. Maybe it's like memory. Someone says they have a bad memory in general (for instance, for names); but it may turn out that they have a good memory for place and direction.

Dowsers often find what are called blind springs where a column of water rises vertically underground. It does not reach the surface, but pushes horizontally along rock fractures as underground streams. These are often associated with places of power, and many European traditional sacred sites are blind spring locations.

Most experienced dowsers are certainly skilled. But for many of us, dowsing is based on intuition. It is not a science, and should be scrutinized by its results, not its repeatability, for as with anything dependent on human frailties, sometimes it works, sometimes not. It is not unlike my experience with homeopathy, which I know can have miraculous results — but sometimes it doesn't. When I drive into the main crowded street in central Bath, I try to visualize a parking space appearing for my car. It manifests often enough to vindicate the method; but when it doesn't, I'm not disheartened.

The Universe, down to the smallest sub-atomic particles, is in a constant state of vibration, at speeds beyond the mind's comprehension. In a living organism, its different organs have different rates of vibration. Animals, trees, stone circles and sacred sites have characteristic vibrational levels. Their common denominator is water, whose subtle energy level one can measure with a biometer called a Bovis scale.

This looks like a semicircular protractor, with a scale going up to 20,000 gigaherz. You measure the dynamic energy level by swinging your pendulum over the centre where the lines intersect, while asking the question 'what is the energy level of this ...' The pendulum should choose to swing up and down one particular line to indicate the appropriate energy level. If the water is in a small container you can try enhancing the subtle energy level with the mind or, as Masaru Emoto does, with the emotive power of certain words (see Chapter 15).

Kinesiology, using muscle testing to discover the body's affinity with chosen substances, is a form of dowsing that illustrates the principle of resonance. To discover whether, say a certain wine, is good for a person or, on the other hand, may produce a toxic reaction, the subject holds the bottle of wine against his abdomen with the left hand, and raises the right arm if he is right handed) to the horizontal. A companion then attempts to push down on the hand, the subject resisting.

If (s)he can't keep his/her arm horizontal, it is because the musculature has lost its integrity. Now the muscles are 70% water. Biological water is in tune with the subject's sense of integrity, which is being challenged by the potential toxicity of what is being presented. This is an instantaneous reaction and is usually very reliable. Many doctors and practitioners use kinesiology for diagnostic testing. If you've never tried muscle testing, why don't you give it a go? It discounts placebo or wishful thinking.

Many experienced dowsers give up the tools, and dowse with their whole bodies. Australian aborigines dowse through the soles of their feet. This is presumably pure intuition at work. The body as a whole organism (like the bee colony) has the insight.

We are mistaken to regard the brain as the source of wisdom. Its role is more like that of a computer designed to cope with the daily demands of survival, rational deduction and other important individual functions. Consciousness exists on two levels; ordinary consciousness is a brain function, while elevated or collective (i.e. non-individual) cosmic consciousness is a function of the whole body, mediated by water. Our biological water, which is connected to the vast, watery network of the wisdom of the cosmos, is the key to our inner knowledge, of which our central nervous system is an integral part.

Cleve Backster's biocommunication research

Cleve Backster is one of the world's experts on polygraph lie detectors and is the originator of the Backster Zone Comparison Test, which is the standard used by lie detection examiners worldwide. He was under contract to develop a polygraph lie detector for the New York City police department in 1966. When a criminal is afraid of being 'unmasked' under questioning, the threat he feels to his wellbeing will make his skin become water saturated (sweating).

This is measured by the galvanometer (part of the polygraph) as a sudden drop in electrical resistance between the electrodes applied to the skin surface, known as the galvanitic skin response, shown by an inked needle on a disc recording chart. Lie detectors work on the principle that when someone feels threatened, or his feeling of safety is compromised, he will respond physiologically in predictable ways.

Backster was not into growing plants. His secretary picked up two at a closing down plant sale. One was a rubber plant, the other a dracaena cane plant about three feet tall. He gave the latter a saturation watering and then wondered how long it would take for the moisture to reach the top. So he attached to the end of a leaf wired clips from the galvanic part of the polygraph — the part that measures skin resistance — to record the drop in resistance as the moisture reached the leaves.

Backster noticed a spike on the recording chart that resembled what was more familiar to him as a human response of emotional disturbance on a lie detector test. He began to think how he could threaten the wellbeing of the plant. First he put a leaf from the rubber plant into a warm cup of coffee. The graph showed a downward trend, almost as though it was bored. Then at 13' 55" of chart time, the thought occurred that he might burn the leaf (he didn't actually *do* it!). The plant's response was instant and the pen went right off the chart.

He then decided to remove the threat by taking the matches back to the secretary's desk. The plant duly calmed down. Backster went into all the possible mechanistic explanations. He was alone — there was no one else in the building. Then he questioned scientists from different disciplines to see if any of them could explain this phenomenon. It was totally foreign to everyone.

This couldn't be called extrasensory perception because plants don't have the human's five senses. This plant's awareness seemed to take place on a much more basic level, thus he coined the term *'primary perception'*. He was so convinced that the plant was responding to his intention that his worldview changed and he decided there and then to make research into this strange phenomenon his first priority, and his experiments continued for the next thirty years.

Once, during the early years, his associate had a wedding anniversary coming up, and his associate's wife wanted Cleve to bring him home to New Jersey on some pretext. They decided to wire up the dracaena plant and kept a stop-watch log of their activities. It turned out be quite a successful experiment. When Backster returned to his Times Square lab that evening, he found that the recorder had noted responses at salient points on their journey: when they went through the subway to the bus station; when they boarded the bus; when it entered the Lincoln tunnel, and at other points during the remaining trip to his friend's home. When they arrived to a houseful of surprise guests, the plant had given a big reaction at exactly the right time.

Backster found that the plant would also respond to a wide variety of microscopic life-forms. Their routine with the electric kettle was to leave it on the cooling hot plate to wait for the next boil-up. This time he needed the kettle for something else, and poured boiling water down the sink. He hadn't planned to do an experiment, but the plant was wired up and responded dramatically to his action. They hadn't done this for months, nor had used chemicals to clean the sink. He wondered whether the plant might have regarded the demise of the bacteria in the drain as a potential threat to itself.

This was confirmed several months later when I took samples out of the sink drain and inspected them under a microscope. There turned out to be a jungle of life-forms present, somewhat similar to the cantina scene from *Star Wars* — all kinds of bizarre life-forms.[3]

He developed a more complex experiment to measure whether the plant would respond to the sudden death of brine shrimps dropped automatically into boiling water at random intervals, while recording their responses at the other end of the lab, and found that they did indeed respond. It was hard to know how to eliminate the energy connection between the experimenter and the plant being tested. He found that if he were tending the plants in any way even for a short time, they would become attuned to him. Plants are territorial. They will be aware of connections between rooms, but not to another unrelated area that may be closer to it.

Then, if he automated the experiment, leaving the lab and setting a time-delay switch for random intervals, making sure that he was unaware of when the experiment started, no matter where he went, the plants remain attuned to him. To start with, Backster and his partner would go to a bar a block away (this, by the way, was in noisy, downtown Manhattan) and, after a time, they suspected that the plants were not responding to the death of the brine shrimps, but to changing levels of the experimenters' excitement.

So the next control was to get someone else to buy the plants and store them in a part of the building that Backster and his friend didn't frequent. On the day of the experiment they fetched the plants, brought them to the strange environment of the lab. They were electroded in the normal way, and wired up to a complex mechanical programmer in a central location. Having been abandoned, the plants would have looked around for something to connect with in their environment. The living shrimps were put in dump cups, which were emptied into the boiling water in random timing. The plants responded to the shrimps' demise with significant regularity.

This classic experiment was written up in the December 1967 *Journal of Parapsychology*. There were predictable condemnations from icons of scientific respectability. At the 1975 American Association for the Advancement of Science meetings, an attempt to discredit him failed. However, in the following few years feature articles appeared in large circulation popular magazines. Backster, who was keen to stimulate interest in his findings, gave 34 lectures to scientific and academic groups and appeared on TV interviews in several countries over the next few years.

After publication in *Electro-Technology* magazine (Dec. 1969), he received five thousand requests from scientists for more information. The publicity that endured was an informative write-up in *The Secret Life of Plants* by Peter Tompkins and Christopher Bird (1973), which became an international bestseller. Quite a few scientists at the time tried to replicate the brine shrimps experiment, but they always seemed to fail because, thought Backster, they misunderstood the need for detachment of mind.

The training of scientists renders them unable to understand the implications of the concept that the consciousness of the experimenter can influence the outcome of the experiment. So Backster found that when they tried to replicate his work with the brine shrimp, they thought they were being rigorous by going into the next room and using closed-circuit TV to watch what happened, instead of waiting to bring the plants in immediately before the experimental run, so as not to allow them to become attuned to the experimenter. They would also do such irrelevant things as washing the leaves with distilled water.

The point is that they were not removing their consciousness from the experiment. It is actually very easy to fail with that experiment. Backster believed that they would probably be relieved to have failed, because to succeed would have gone against some widely believed scientific 'laws'. However, the insistence of repeatability as the criterion for scientific proof is actually anti-life, as life itself is not predicable. Repeatability is about control, which is the hallmark of western science and culture.

Not only is spontaneity important, but also so is intent, as he had discovered that first day. If you pretend, it just won't happen. If you say you are going to burn the leaf and don't mean it, nothing will happen. Trying to repeat an experiment the same way won't work either. Repeatability is a feature of machines, not living systems. Backster tells his students:

> Don't *do* anything. Go about your work. Keep notes, so later you can tell what you were doing at specific times, and then transfer them to your chart recording of tracing changes. But don't plan anything, or the experiment won't work.

Backster usually found a new line of research by accident. Working late at night, he was stirring the jam at the bottom of his strawberry yoghurt carton when he noticed an unusual reaction on the plant's chart. He wondered if he might get some interesting results if he electroded live yoghurt cells. This turned out be some of his most productive research, for the bacteria in yoghurt seem to be particularly attuned to human interaction in their immediate neighbourhood.

It was challenging to design appropriate equipment. The bacteria were so sensitive that it was necessary to block the view of the chart recording to increase the spontaneity of conversations. Even observing an experiment can influence the result. Backster discovered that once you allow your consciousness to interact with an experiment, it could affect the outcome, so he designed a new experimental technique that randomized the timing, allowing him to withdraw from the experiment itself.

In 1979 the Backster School of Lie Detection, now located in San Diego, gave a reception on two floors of the building to hundreds of people attending a polygraph conference. There was a bar on each floor. Before the crowd arrived, Backster electroded some plain yoghurt, expecting chaotic readings because of all the commotion. Usually they conducted experiments in a controlled environment, but he wanted to demonstrate only the receptivity of the bacteria.

Having found that bacteria were particularly sensitive to local human behaviour, here was an example of selectivity. They were much more 'interested' in fellow bacteria being killed by the alcohol than by the human activity. This ability to prioritize was an example of a primary perception process similar to human psychology, but on a bacterial level.

Elisabet Sahtouris, an evolutionary biologist has said:

> Bacteria are responsible for forming the larger cells from which all other life kingdoms are constituted. Further, bacteria are the only creatures that could survive without all the others. Why should bacteria not think, if they could think, that the world is all theirs?[4]

A cytologist, Dr Howard Miller, in a review of Backster's research in *Medical World News* predicted that Backster 'may have discovered a kind of cellular consciousness'. (See for instance, Figure 20.)

Having successfully gained meaningful evidence that suggested the possibility of biocommunication in plants, shrimps, chickens and a variety of bacteria, Backster wondered whether this might be obtained at the cellular level in humans.

Studying in-body *(in vivo)* cells, he thought might involve too many complexities with nervous system activity. So he decided to find a way of extracting living cells *(in vitro)* in order to see if, when electroded, they might communicate with their host organism. Research with blood would require medical supervision. However, a dental researcher he'd met had developed a method of obtaining white blood cells (leukocytes) by centrifuging cells obtained from the donor's mouth,

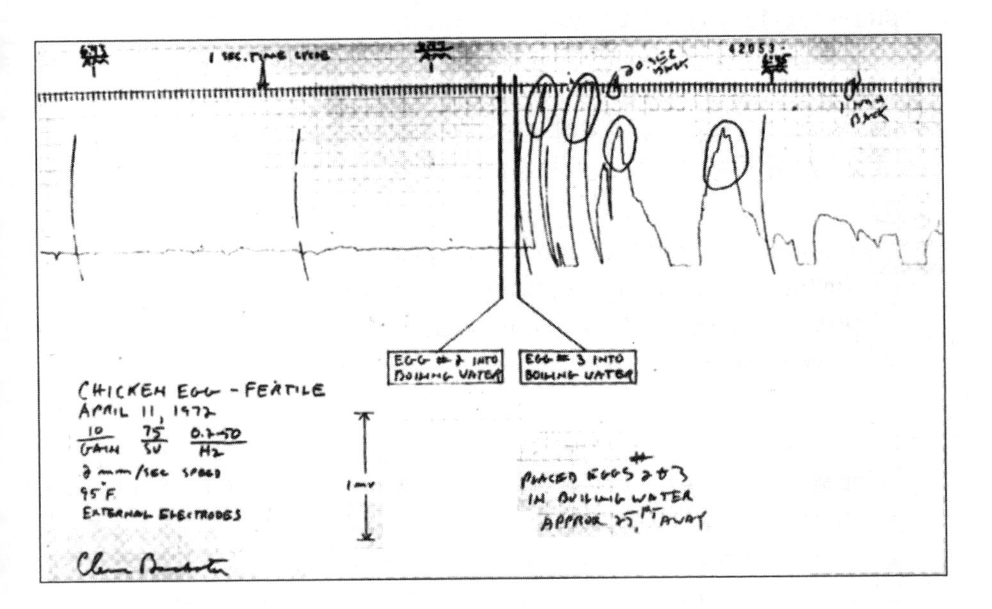

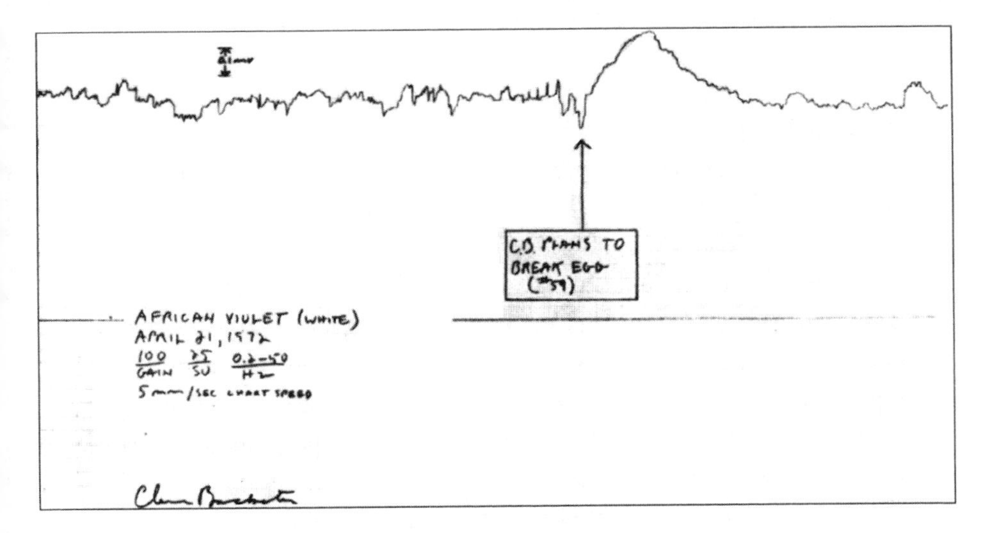

Figure 20. Cleve Backster's findings. (a) Chicken egg monitoring other eggs: An egg wired-up to a polygraph galvanometer (lie detector) shows great distress as other neighbouring eggs are plunged into boiling water. (b) The violet and the chicken's egg. Backster wired up an African violet to a polygraph galvanometer (lie detector). It recorded the moment when he planned to break an egg. (Cleve Backster)

which would not require medical supervision. (He describes these procedures in his book *Primary Perception.)*

The most extraordinary test done with swabs from a person's mouth showed that, when the parent organism (a doctor) was several hundred miles away, the cells were still 'conscious' of stressful events in her life. As Backster says:

> Sentience does not seem to stop at the cellular level. It may go down to the molecular and beyond. All sorts of things that have conventionally been considered to be inanimate may have to be re-evaluated.

It is one thing to say that repeatability is a feature of machines, not living systems, which seem to be more spontaneous. Chaos theory gives a glimpse into a different world. Backster's experiments showed that you can't fool the plants by pretending. A lot of conventional human behaviour is insincere or hypocritical — this is accepted as normal. At the etheric level, this doesn't seem to work. If our human consciousness was at the etheric level, is it possible that we would be altogether more transparent?

There has been really no support for his research in the USA or in Western Europe. On the other hand, in Eastern Europe, Russia and Eastern countries, where psychological, emotional and spiritual factors are more accepted, Backster's research is celebrated as pioneering.

Backster has received only disinterest or scorn from US scientists. The Western scientific model requires repeatability, predictability and control to validate any experiment. This emphasis on repeatability is however, anti-life, in the sense that life's experiences are inherently non-repeatable. Science has not yet accepted that the attitude, emotional state and bias of the researcher can affect the outcome of an experiment; that mind can influence results.

There can be no such thing as truly objective research unless self-consciousness gets out of the way. Backster arranged randomized triggers to stimulate responses in the plant or culture to ensure that the researcher was not influencing these responses. Our current worldview insists that only humans experience consciousness, and that it is a function of the brain. This research disproves both of these assumptions.

It would be hard to overestimate the importance of this research, for it goes to the heart of the basis of life — that all of life is interconnected and interdependent, and that all actions have wider consequences. It lifts a little the edge of the intricate web of life.

There seems to be a consciousness at the heart of Nature that may be the engine for a higher meaning and purpose to do with evolution; and the evidence is that *water* itself is the vehicle of this consciousness. Water (or as blood and sap), the main constituent of all living things is, as Viktor Schauberger pointed out, the carrier and fountain of life itself.

Backster's observations showed that the signal could travel instantaneously dozens, even hundreds of miles. To test whether it might be electromagnetic, he placed the plant in a Faraday cage or in a lead container, the normal screening devices, but they made no difference to the plant's receptivity. He was convinced that it did not fall within the electromagnetic spectrum.[5] He did not consider water as a communication channel common to all life-forms, cellular and molecular, for such a weird possibility was then not on the radar. It now remains a strong possibility, but how?[6]

Quantum entanglement

When Cleve Backster wrote up his research, there were no explanatory models to turn to for how biocommunication could work. It clearly must have a quantum quality about it, and it occurs to me that the new theory of quantum entanglement, which we discussed earlier in the context of Mae-Wan Ho's experiments, could throw some light on it (see Chapter 11).

Quantum entanglement is a mathematical description of what happens between a pair of particles. In a typical experiment, a light particle, called a photon, is made to split into two (daughters) to observe how they behave. They are not pre-programmed, but it is found that they respond spontaneously, independently, and in an interconnected manner.

It is the act of observation that makes them appear, and time and location seem to be completely irrelevant. It seems that we could use pairs of entangled particles for an instant communication system, perhaps in a manner similar to what Mae-Wan Ho describes with chains of collagen fibres in the human body, for what she calls 'quantum coherence' in the organism is the next stage up from 'quantum entanglement'.

There are more complex conditions that govern the existence of particles. The first is that their properties do not exist apart from the act of observation; they are aspects of mathematical measurement. Secondly, special and temporal considerations do not enter into the calculations; the particles could be anywhere. Lastly, the existence of the quantum field ensures that everything is entangled, so very special conditions need to exist to produce particles that are not entangled.

This could be an explanation for telepathy or for paranormal experiences; and why not for empathy? There is a new branch of physics called 'Quantum Information Theory' which looks into how information within organisms can be transmitted, by a mixture of entanglement and classical information transfer. This is an exciting new frontier.

Mathematician Chris Clarke sees the world as a nested lattice of quantum organisms. There is a constant interplay between the coherence that each system receives from the greater whole. He calls this 'Living in Coherence', and it may be a metaphor for the influence of a supreme intelligence:

> This suggests that the world, rather than being a collection of
> isolated particles pushing each other around,
> is more like an intricate web of subtle interconnections.[7]

If we can conceive of water as a vehicle of consciousness, it may not be such a large step to consider whether water may have a memory.

15. The Memory of Water

'The water will tell you,' said the guide, when the travellers asked him how deep was the water.
Plato, *Theaetetus*

Information transfer

Ecological theory teaches us that all life is interconnected, interdependent. The idea of sustainability follows Nature's way of conserving energy — as circles within circles, nothing is ever wasted. At the end of its term it is recycled and contributes again to the whole with a new role — in a vibrant system, perhaps a more purposeful one. Nature employs closed systems with organisms to promote enhanced evolutionary change and growth. This allows energy to be conserved, raised and reused.

A machine, on the other hand, is an open system. It obeys the Second Law of Thermodynamics, which says that every system runs down and entropy or energy loss results. A motor car, no matter how well designed and constructed, is a machine that will wear out with constant use. If our bodies were machines, in accordance with Newtonian physics, they should wear out in a year or two. The fact that our bodies can last seventy years, or in the best possible conditions, even twice that time, shows that they have properties of a closed system, conserving energy.

Mainstream science recognizes many of water's qualities, including the anomalous: it is a powerful solvent; its versatility; its need to move and pulsate; its property of balancing heat. However, because of a materialistic viewpoint, mainstream science does not comprehend the energetic or quantum aspects of water: information storage and transmission; how it can self-organize, self-purify; how it exhibits some of the properties of an organism.

All living cells, tissues, organs and organisms have their own fields. The networks of living systems are regulated by the information clusters that the morphic fields create through their living memory bank, which is connected to the collective memory banks of similar objects or organisms. The enrichment of this memory, according to Rupert Sheldrake, is what drives evolution.

Nature tends to convey information as a repetitive signal. So the repeating geometric form of the fractal is what Nature uses to relate the macro world to the micro, through morphic resonance which influences both form and behaviour. Sheldrake's research shows that it is easier for humans and animals to learn a new skill that other humans/animals have already learned (the 'hundredth monkey' syndrome).

Fractals

Fractal geometry was born out of chaos science, which demonstrates that the natural order is not random, because out of chaos there inevitably emerges an infinite order. Everything in Nature is organized into repeating geometrical forms. If we look at the irregular shape of a coastline, we find that its particular shape of irregularity is constant at every scale. The world is regular in its irregularity.

Fractals describe the arrangement of beach sand, a river's tributaries, a lightning strike, the lines a palmist studies, the formation of clouds, the bronchia of the lung. Fractals are generated by the quantum field working with water in the transfer of information. As we saw in the study of vortices, turbulence or chaos in water provides the opportunity for the transfer of high quality subtle energy or information.

Jacques Benveniste (see below) was one of the first to discover that water could be biologically active in the absence of solutes — which contradicts a fundamental biological law: namely that biological activity in water requires a significant amount of that substance in solution.

About the time Benveniste was publishing accounts of his research with high dilutions of biologically active chemicals, Emilio Del Giudice and Guiliano Preparata, physicists at the University of Milan, were

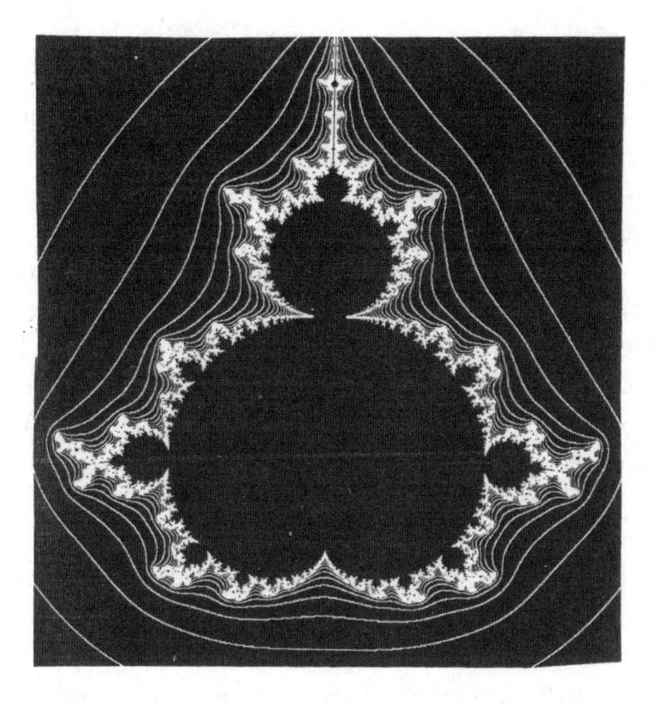

Figure 21. The Mandelbrot fractal. The Mandelbrot Set is the best known of several mathematical fractal models which describes how natural things grow. This helps us better to understand the development of unpredictables like weather systems, natural turbulence, wave development and the growth of plants. The sets are created on a computer by applying numbers to a plane according to a given formula, repetitively, at speed. The boundary area just beyond the black shape oscillates continuously producing magical shapes and colours, the basic pattern of the set repeating itself with successive enlargements. (A. Bartholomew)

working on a model of molecular vibrations in water — 'the theory of electrodynamic coherence'. They postulated a magnetic memory built by means of coherent molecular vibrations (water molecules pulsating in phase with their magnetic field).

Biological water is influenced, either beneficially or harmfully, by electromagnetic fields (through metals and minerals), by biological fields (in organic substances) and by coherent domains, for all of which the vehicle is water.

Martin Chaplin (see Chapter 11), in his two-states of water research, discovered a change in the electromagnetic qualities of the less dense (non-coherent) phase — the atoms of different polarity repelled each other — while at dense (coherent) phase they clustered closely together, regardless of their conflicting polarities.

According to quantum theory, every particle is connected or 'entangled' with every other particle at every scale from micro to macro. These connections within an organism or between organisms are based on the quantum coherence of water-based life.

Life is possible only because the quantum qualities of water allow it to transcend its own chemistry. The key seems to be that the hydrogen bonds between molecules are ten times weaker than the hydrogen-oxygen bonding, which enables water's structures to resist compression, vibrating continuously — bonding and breaking in order to retain its energy and cohesion.

What happens with DNA is even more remarkable. It has the densest, most complex structure imaginable in order to synthesize proteins. When a protein approaches the DNA, the water molecules surrounding it stop vibrating. The specific protein structure causes the interfacial water to break the DNA strand at an appropriate point in order to allow the transfer of information from the protein to the DNA — an extraordinarily sophisticated process.[1]

Quantum phenomena can be quite fickle. Light can behave either as a collection of particles or as a wave — depending on the observer's point of view. This makes reductionist scientists wary, just as they are with the placebo effect in medicine (even though it is an example of one of the most creative aspects of the human mind).

Memory or recording?

Much has been written about water's memory. Does water have a memory? The suggestion is confusing, for memory is a considered to be a brain-based activity, many memories being stored, which may be accessed later. This is rather like the discussion we had in the preceding chapter about the receptivity of plants.

I've heard of people claiming that memory can be inherited. A common example is of a daughter who insists that she has the exact dreams that her mother used to have years ago. Is it possible that a family's biological water may play a part in this?

When it is moving, water is sensitive and records vibrations. At some future time we will learn how these are discriminated, perhaps as a complex spectrum of resonances of the kind used in vibrational healing techniques. These therapies recognize that each organ of the body (some even say each molecule) has its own resonant frequency. They are like numbers on the atomic scale, but for energies.

Water records qualities and resonances, not memories as such. The great Antarctic ice sheet contains physical records, like dust, pollen and air bubbles, which can tell us things about ancient climates. Perhaps the ice also records resonant information relating to qualities.

This means that the medium of greater coherence — the etheric — is able, under certain circumstances, as with the river's immune system, to raise the water's dynamic energy level. Viktor Schauberger's technological applications were based on water's abilities to absorb, retain and enhance energies. The problem for mainstream science is that these energy qualities are not easily quantifiable; they are hard to measure.

Photons and water

There are subtle energies connected with water that are hard to measure. One of the controversial areas of energy study is photon research. Photons were discovered only in the last couple of decades, with the development of dark field photo-microscopy, and are usually interpreted as a sign of life force. They appear as bright spots of light against a dark field. Some go as far as to say that: *energy = light.*

Unhealthy blood will show low photon presence, while a sample of blood from a healthy person will show an abundance of photon light spots. This analytical method should revolutionize medical symptom research, but unfortunately it is still met with much scepticism. One of the pioneers of this research is David Schweitzer, known particularly for his research into blood. He analysed different samples of water in his dark field microscope demonstrating that the quantity of photons indicated the vitality, and probably the quality, of the water sample. (See Plates 19–21.)

Schweitzer found that blood cells have an esoteric geometric structure and harmonious colours. In 1996 he discovered how to photograph the stored frequencies in homeopathic remedies and to record the influence of positive or negative thoughts on bodily fluids.

As good quality, living water is vital for our health, we need to be able to discriminate between the many who claim that their water treatment product is the best.

Storage of information

When water is flowing as its nature dictates, energetically in spirals and vortices, it creates the structure necessary for it to carry constructive information. These are microclusters of vibrating dynamic energy centres, constantly receiving and transmuting energy from every contact the water body makes. Despite water's fluidity and its ability constantly to change its state, the molecules, if conditions permit, generally organize themselves into structures.

The vortical movement creates the polymer liquid crystalline chains and the laminar structure that generates dynamic energy from the interaction of their plane surfaces against each other. These structures can be observed with a suitable microscope. The more powerful the vortical action, the greater the storage capacity of information (like adding 'memory' to your computer). Thus water put through Viktor Schauberger's implosion (powerfully vorticized) process can enhance the energy of organisms which imbibe it.

The clusters have the ability to store vibrational impressions or imprints. If these are beneficial, they may be able to restore healthy resonance in the human body, as through homeopathy. On the other hand if they are the imprints of toxins or pollutants in the drinking water, they may be carriers of disharmony and disease.

We shall now look at some examples of research into water's 'memory'.

Jacques Benveniste and homeopathy

Jacques Benveniste (1946–2005) was a medical doctor with a particular interest in immunology. Working at the French National Institute for Health and Medical Research (INSERM), he studied the effect of extremely weak solutions of antigens (compounds that elicit antibody responses) on basophils. Basophils are special cells in the bloodstream that contain tiny granules of histamine, the substance released in classic allergic reactions like hay fever. The antidote is to take anti-histamines.

When antigens bind to special receptors on the surface of basophils, they trigger a reaction that causes the basophil to degranulate, and to release histamine. According to classical Newtonian theory, progressively larger amounts of antigen should produce progressively stronger cellular reactions; and more dilute solutions of antigen should produce weaker cellular responses. Beyond a certain dilution there should be no reaction at all, simply because there would, statistically, be not even one molecule of antigen in the test sample.

Contrary to established scientific theories, as thus greatly to his surprise, Benveniste discovered that ultra-weak solutions of anti-immunoglobulin E (algE), containing no molecules of antigen were still able to induce significant basophil reactions, causing cells to degranulate and release their antihistamine.

In homeopathy, common practice is to shake a remedy vigorously in water, take one tenth of that and add nine parts pure water. Ten such dilutions are called a decimal. After 12 decimals there are no molecules left of the original substance. Nevertheless the remedy still has biological activity. Increased dilution can make the remedy even more potent. How is this possible?

Benveniste knew that his research results were so controversial that his only chance of publication was to have other scientists duplicate his work. After his experiment was replicated experimentally in several laboratories around the world, Benveniste was finally able to have his article accepted by the influential magazine, *Nature*. This was accompanied by an editorial questioning whether this research was to be believed, as it contradicted current scientific theory.

Nature had thus published an article that seemed to validate the principle of homeopathy, a healing practice considered to be bogus by most of its readership. The resulting worldwide storm of controversy was like a battle between old-world Newtonian thinkers and new-world Einsteinian theorists and clinicians, each appearing to be living on a different planet.

The reactions to the publication were brutal. Within four days, *Nature*'s editor, John Maddox, descended on Benveniste's lab in Paris, bringing both a physicist specializing in scientific fraud and a professional magician, known for his work in 'debunking psychics and paranormal phenomena'.

They stayed for five days, labelled successful experiments as 'fraud' and Maddox tried to repeat the experiment himself without the protocols. Shortly after their return, Maddox published a damning report in *Nature*: 'High dilution experiments a delusion'.

What the debunking team failed to understand was that these experiments were to measure subtle energy effects. There was no consideration given to the possibility of the researchers' attitude influencing the outcome of the experiments.

Indeed, later Benveniste so despaired of laboratories trying, but failing, to reproduce his results that, to ensure that the method was standardized, he automated the experiments with what he called a 'robot'. Even so, there seemed to be the odd researcher in a couple of laboratories who could not succeed. Eventually he traced this down to the researcher emitting a blocking signal; one of them was a frequency scrambler.

In his final years, Benveniste investigated the possibility of recording electromagnetic signals in biological water. Life depends on these signals between molecules. He recorded digitally the waveforms of some thirty substances (for instance, bacteria, antibodies, and so on) on to a computer hard drive. These electromagnetic reproductions produced the same reactions as the original substances.

He describes observations that cannot be explained by current theories. 'The Benveniste Affair' is notable because it shows the extent to which the scientific establishment will go to discredit anyone who challenges cherished theories, even going against well-founded ideals of open-mindedness. The fall-back position is: 'We can't explain it, so it can't be true'.

The question is: How can widely spaced molecules communicate with each other, and how can water store information? One of the anomalies of quantum water is its high dielectric value, which allows large electric fields within living cells.

Quantum mechanics discovered that packed molecules go from a chaotic to a coherent state if that ordered state contains less dynamic energy. According to Preparata and Del Giudice, stability can be achieved only through the existence of long range energy forces, which lead to collective behaviour. This is called the Theory of Coherent

Domains, an example of holism (as against discreteness).

Just before the visit of the *Nature* fraud squad, Dr Atrias, a homeopathic doctor, had persuaded Benveniste to try out an electrical machine which he claimed transmitted chemical information. It was found possible, purely by means of this machine, to transmit the information from a solution of algE potentized to 30 decimals into a glass tube containing a dummy dilution. Benveniste had for long suspected that the electromagnetic properties of the cell played a part, that by this means the information was stored in the water before the molecule disappeared, and then was able to be recalled at a later time.

While it certainly appeared that Benveniste's team were able successfully to demonstrate that a chemical signal could be transmitted without its original molecular support, it now looked as though an electromagnetic image of the original molecule could be held. He also confirmed that without the medium of water, signals cannot be transmitted in the body.

Homeopathy works on the coherence of the whole organism, so you would expect it to respond to quantum rather than material principles. One of the issues with testing homeopathic reactions is to discover methods appropriate to the individual person. Normal detection methods apply to examining individual molecules, rather than the collective global properties of an organism.[2]

We may still be far from understanding just how water records information in its structured clusters, but it is undoubtedly tied up with water's extraordinary properties.

Masaru Emoto's research

In 1986, Dr Masaru Emoto was sourcing equipment for acupuncturists and moxibustion practitioners. In particular he was looking for new low frequency therapy equipment for his Japanese health products agency. He was concerned that while the technology existed for measuring water's content, there was none to measure its quality.

He asked Dr Lee Lorenzen, who was later to develop microcluster water, if there was such technology in the USA. Lorenzen sent him three Bio-cellular Analysers (Magnetic Resonance Analyzers that are used in homeopathy for measuring levels of vibrational energy), but

Emoto couldn't make them work. When he later lost his business, Masaru thought he would experiment with those three MRAs in the basement. Eventually, with one of these machines he was able to produce high quality water which he called Hado water, that he found when he gave it to his staff's families, produced healing effects.

Hado is a magnetic resonance pattern of wave motion measured as a vibrational rate, known in Chinese tradition as *chi*, a vital energy originating 'from the circulation of electrons around the atomic nucleus'.[3] Emoto believes that the hado changes according to the consciousness of the observer. (Since 2005 he has used the term to mean all the subtle energy that exists in the Universe.)

Emoto found in a book on 'mysteries in science' that 'there are no two snowflakes that are alike', and he realized that taking pictures of the water before and after treatment with hado (MRA treated) — frozen as crystals — would show how information can be imprinted on water. Ice crystals produce a time-freeze of vibrational energy.

The technique for producing good crystals from water samples proved to be tricky, but ultimately he and a colleague succeeded. It involves putting drops of the water to be sampled into 50 petri dishes and frozen for 3 hours at -25°C *(-13°F)*. They then warm up in the lab where the temperature is near to freezing. At about -15°C *(-5°F)* a crystal may form, but only for some 20 seconds, so it must be quickly photographed under the microscope.

Each water sample will produce mixed results: if the water is of good quality, some crystals will be well formed, others poorly, and some of the ice grains remain amorphous. Apparently, though, it is possible to find a common theme amongst the better formed crystals (none of which will be identical). Their interpretation must be more art than science.

This cannot be regarded as a scientific method, as the results are variable and unpredictable (and subjective?), but this does not seem to worry Dr Emoto! However, this aspect of his method means that he is unlikely to get mainstream support for his research and that other researchers would find difficulty in replicating his experiments.

He started testing many sources of drinking water, finding that most municipal supplies produced a poor crystalline structure — hardly surprising as they would contain chlorine. The few that did make crystals he found had significant proportions of groundwater. Sacred springs, like Lourdes, produced the finest crystals. (See Plates 10, 11, 12.)

Music and resonance

One of Emoto's assistants suggested exposing water to music. This proved to be very successful. A bottle of distilled water was placed between two loudspeakers to produce samples. It soon became apparent that works of the great composers like Bach, Mozart and Beethoven produced beautiful crystals, whereas those formed after water's exposure to pop music like 'heavy metal', were ugly and ill-formed. The great works are composed according to natural laws of harmony, which would show up as resonance in the crystal structures. (See Plates 13, 14.)

Resonance occurs between two systems when their vibrational rates are in harmony or at a critical point of disharmony. It can be associated with both creative and destructive phenomena. You've probably heard that a specific musical note can shatter a fragile wine glass!

It is possible that resonance operates both through the water medium and through the etheric medium. A relationship between human beings or between a human and an animal can be seen as an emotional resonance through the medium of water. On the other hand, when you resonate with a particular gemstone or mineral, that would be through the medium of the etheric.

David Tame believes that the degeneration of popular music has often come before the fall of great civilizations.[4] Dr John Diamond describes how certain types of heavy metal and rock music, with its 'stopped' quality at the end of every bar, weaken the body's muscle tone and leads to decreased performance of children in school.[5] The more flowing natural rhythm corresponds more to the heart rhythm, supporting a natural balance in the body.

The human body's high biological water content means that we respond deeply to different kinds of sacred music. It should be possible to calculate mathematically the geometrical structure of a Bach fugue, just as has been done with Rembrandt paintings.

Dr John Ott asserts that the body has its own natural resonant frequency that can be supported, over-stimulated or suppressed by different light, sounds, or electromagnetic frequencies and vibrations.[6] The long term constant bombardment by electromagnetic emissions from high tension electricity cables, TV, radar, microwave and transmitters has a significant effect on our overall health and collective behaviour.

Water as a medium is very sensitive to resonance. Its electromagnetic qualities make it the ideal path for the intercommunication of resonant frequencies. Ralf Roessner, who developed the double-egg water vortexer, noticed that a change also occurred in his own bodily fluids — the 'instrument' of the whirling, observing person — when he holds the egg vortexer against his body.

Language and water

Masaru Emoto had the idea to see if words taped to a sample bottle would carry dynamic energy into the water. Indeed they apparently did — in whatever language they were written! The comparisons between phrases like 'I love you!' and 'I hate you' produced the anticipated results. He insists that words alone carry a subtle energy, which may puzzle some who would prefer to believe that the words carry the consciousness behind their creator. What is the sensitive person picking up who holds a book declaring 'this book has good vibrations!'? It is surely the insight (subtle energy level) of its author, not just the words.

Emoto uses his machines for healing in a way similar to the way radionic practitioners and sound healers do in the West. He discovered that the quality of water improves or deteriorates according to the information to which it is exposed. He deduces from this that, as we are composed principally of water, we are likely to be more healthy when we receive good information (energy), and suffer when exposed to bad energy. There are clearly many more factors involved, but you can see what he is getting at.

Perhaps the most dramatic research Emoto directed was to organize large groups of pilgrims to pray by lakes in order to help purify the quality of the water.

Masaru Emoto's philosophy

As Masaru Emoto is an inspired and successful speaker with millions of supporters, it might be helpful to summarize his philosophy:

—Emoto's message is concerned with energy vibrations. Quantum mechanics acknowledges that substance is vibration. Emoto describes how, at the microscopic level, it is the number and shape of the electrons orbiting the nucleus of the atom that give each substance a particular set of vibrational frequencies. Whatever the object, nothing is solid; the appearance of solidity is the nucleus surrounded by a wave rotating endlessly at great speed.

—Each human individual vibrates at a unique frequency, depending on their energetic state of love, joy, or of negative intention. He says the same principles apply to objects and locations. For example, there are places where accidents seem to happen, locations where enterprises are successful and objects that seem to bring tragedy to successive owners. Intense subtle energy can produce natural disasters.

—When water freezes, the particles link together to form a crystal nucleus. If the nucleus grows coherently, a visible water crystal appears, but when information is present that is inharmonious to Nature, an incomplete crystal results.

—Everything in a state of vibration also emits sound. This is why crystal formations are affected by music, and the subtle energy of specific words can affect the quality of crystal formation. Water can convert these vibrations into a pattern visible to the human eye.

—Emoto believes that in harmonious Nature only vibrations of love and gratitude are present — the trees and plants show respect for each other, and animals take only as much food as they require. He believes that humans learned words and language from Nature. He says that words like 'gratitude' and 'love' form the fundamental laws of Nature and the phenomena of life. However, words like 'you fool' do not exist in Nature. Such words and thoughts are created by humans.

—Natural water sources produce beautiful crystals, but much of Earth's water, whether in the atmosphere, on the surface or even underground, is polluted. He says that pollution originates in human consciousness; it is the creation of our selfishness and our egos.

—Emoto claims that human beings vibrate 570 trillion times a second. The human holds a universe within filled with overlapping frequencies, which produce a symphony of cosmic proportions.[7] When one being creates a frequency and another responds with the same sound, they resonate. People who generate similar frequencies attract each other. When their frequencies are incompatible, they cannot resonate and people cannot accept each other.

—Our emotions affect the world moment by moment. If you send out harmonious thoughts, you will help create a beautiful world. The key to what is possible for you lies in your heart to know, and your will to make happen. What we imagine in our minds becomes our world.[8]

—Our state of consciousness certainly affects the world around us. If you bake a cake lovingly, some will note that it tastes better. I know others will scoff at these claims, but most readers will know what I mean, even though it is hard to prove this scientifically.

—Emoto notes that a common thread of the great teachings of the world's religions is that intentional thought can change the quality of what we experience. This can be practised through the meditative technique called vizualization. We all have the power to vizualize a protective shield for self or others; a state of becoming more healthy, a homeopathic remedy. A champion tennis player cannot succeed without a powerful ability to visualize his best shots. The power of the mind to affect outcomes is really extraordinary, but we need to exercise it with careful judgment and humility.[9]

—Most objects in Nature emit stable frequencies. Each sparrow sounds basically similar, and the sounds made by dogs and cats do not vary much. But humans are able to use a full scale to make beautiful melodies.

— Emoto says that humans are the only creatures who can resonate with all other creatures and objects in Nature. We can give out energy and receive it in return. This is why it is vital that we change our way of thinking so that we can live in harmony with Nature. If you fill your heart with love and gratitude, your life will give you what you seek. But if you emit signals of dissatisfaction and hate, your life will become this.

Masaru Emoto's research is very important in motivating people. His books are international bestsellers, and he is in great demand as a lecturer (albeit usually through a translator). Water is neutral, but our terrible abuse of the Earth has turned it into a carrier of disease. Emoto shows us how water can also be a medium of healing. He is passionate about his mission. This is shown best in the third volume of his photographic albums, which deals with the influence of the energy of positive thinking and of prayer.

The fact that his albums have sold significant quantities in about eight different language editions demonstrate the widespread desire to believe in the spiritual nature of water. There's nothing more convincing than a photograph. Emoto is fulfilling an important role.

If we were really to take on board that we can heal Earth's groundwater, I believe that Nature would respond, for the groundwater is the key to the health of Gaia. It could happen quickly, given the will. We caused the pollution; we can clean it up by being the channels of its healing. In this way we can demonstrate our responsibility as Earth guardians.

Andreas Schulz's water crystal research

Andreas Schulz's water crystal analysis is a method of assessing the quality of a water sample, without adding anything to it, which attracted a lot of interest when it was introduced in 1993, at an exhibition in Freiburg, Germany. It is easily understood by the non-scientist, and is surprisingly simple, not requiring particularly complex equipment.

The water sample is first distilled at a relatively low temperature (70–80°C). Part of this is desiccated into a powdery residue which is combined with some of the distilled liquid, placed on a glass slide and allowed to crystallize at room temperature. The resulting crystals are photographed through an electron microscope.

Compared to Masaru Emoto's analytical method, the quality of the water sample is determined by a standardized protocol. For example, the crystals will form particular angles. It was found that the more star-shaped crystals and 60° angles that appear, the higher was the quality of the water. A tendency towards 90° angles indicated low energy and polluted water. If the 90° structures predominate, the water is not suitable for drinking. Other criteria include the strength and spread of the crystal formations, fields of darkness, border structures, etc.

In conventional crystal research, the microcrystalline structure of substances is dependent on the mineral composition of the sample, showing, for example, rhombic, tetrahedral or hexagonal forms. Such information can help assess the material characteristics such as strength and load capacity. This method, however, will not help determine the quality of a sample of food, for example, as to its level of toxicity, or its ability to improve the quality of life.

Schulz's method of crystal analysis depends on macrocystalline (larger) structures, shapes which are independent of the chemical composition. So, while the conventional microcrystalline structure will show sea water with cubic salt crystals, the macrocystalline sample of high energy sea water will tend to show rounded shapes with different angular structures. As this method is more specific and more reliable than Emoto's, it has achieved wide scientific acceptance. It has been shown to have a high level of precision.

Chemically pure water passing the physical tests may not be of high quality. The consumer is often confused by the technical language of water analyses made by municipal or commercial water companies which rely only on the physical analysis. The value of the Schulz method is that the energy level of the samples which are paramount for quality is easily seen.

What contributes to the overall picture are the inner structure of water, the arrangement of cluster structures, the total information content and the state of the water's vital energy, which will be affected by any form of energy pollution. This makes it a more reliable measure of water sample quality not only for drinking, but for evaluation of the quality of food, medicines, cosmetics and a range of products. (See Plates 22–25.)

PART 3

Facing the Water Crisis

16. How We Treat Water

But there is not, as they say, any worse water than water that sleeps. (Mais il n'est, comme on dit, pire eau que l'eau qui dort.)
Molière, *Tartuffe*

Except for floods and droughts, we ignore water. In Europe and most developed countries we seldom need to think about it. It's there when we turn on our taps, and then it drains away somewhere else. We can bathe when we choose. In summer we water our lawns on a whim; the fountains in the parks keep flowing.

Most thoughtful people accept that our culture is at war with the environment. The Hopi shaman would be horrified that we are extracting uranium from the Earth in order to make energy to drive motors. He knew its important Earth energy balancing role in the ground.[1] But how many realize that our desecration of fresh water is even more damaging to Earth? Water is much more than a 'resource'. We have suggested it is the handmaiden of the quantum field in guiding life and evolution. As we are beginning to see, water is the essence of life itself (see Chapter 19).

You can't take good health for granted. Nor should you water. Water supplies are now gravely under threat. We don't know how to manage or take care of fresh water. We waste it prodigally. In California, for example, 78% of the water supplies are used for poorly managed agricultural irrigation and only 22% for industrial, domestic and environmental use. It can take a thousand gallons of fresh water to produce an eight-ounce steak. The United States uses three times as much fresh water per day as any European country, and enormously more water than any developing country.[2]

Climate change is bringing redistribution of (often torrential) rainfall and increasing drought. These unfamiliar conditions will force us to become much more careful with water. Sadly, they will also without doubt bring conflict between nations in their search for the very staff of life.

Our virtual world

Global warming was blamed for the devastating floods in England in July 2007. While some local authorities had made preparations for a flood of this proportion (for instance, by building collection basins), and fared better as a result, we need to look much deeper, to the heart of our problem with water. Climate change may indeed have played a part in their unpredictability, but there is another cause that we urgently need to recognize. We have forgotten what our forefathers understood — that natural land is like a giant sponge, which can soak up most of the deluges that Nature can inflict on the environment.

Nearly all the major floods of recent decades, Mozambique, Columbia, Costa Rica, Bangladesh, were due to the heedless destruction of the most important part of that protective sponge — the forests of the watershed. The denuded land, exposed to the sun, sheds the downpours like water off a duck's back, for the rain will soak into the ground only if the surface is cooler than the rain.

It is tragic how little understanding there is of the importance of tree cover to prevent soil erosion by increasingly violent storms. The terrible floods in Haiti in the wake of 2009 hurricanes was due to deforestation — caused by relentless population growth, dependence on charcoal as fuel and building on the floodplain.

Our culture is completely ignorant of the importance of healthy river systems to keep the environment in healthy balance. Excessive deforestation clogs up the rivers with silt and loss of riparian vegetation causes overheating and energy loss from the rivers. The Ganges and Yellow Rivers are almost stagnant from siltation and upstream water extraction and the Ganges from faecal coliform pollution.[3]

Those who lived on the land had the knowledge of how to care for it passed on by those generations who came before. The farmers knew their land intimately; how important it was to keep the soil, woodland, the bogs and wetland in good state. The way they farmed kept the soil healthy, with its countless species of bacteria, microbes and worms. A healthy soil is a good sponge. Also, builders used to have the wisdom to keep off flood meadows, the land that, willy-nilly, is likely to be subject to periodic flooding.

Modern farming practices are usually neither eco-friendly nor sustainable. In industrial farming the land is used as an anchor for plants that are chemically fertilized and weed controlled, and often watered by wasteful irrigation. The chemicals kill off many of the microbes and worms, damaging the soil structure, making it less able to retain moisture and reducing its nutritional value. Their over-deep ploughing techniques use steel blades that put an alien positive electric charge into the soil, which encourages pests that have to be dealt with by more chemicals.

This is virtual farming by virtual farmers many of whom, ranging around their vast acreages in enormous air-conditioned, radio-equipped tractor cabs, have lost contact with Nature and even with their own land. They have lost the plot, seduced by technology and by their power to subdue Nature.

As we shall see, water is the driver of all natural processes in Nature. Our materialistic worldview and an obsession to get rich have been the cause of the destruction of the equatorial rainforests, the vital mechanism for moderating world climate. With no respect for Nature's laws, the modern technologist, fuelled by cheap oil, greedily exploits the Earth's resources, carving great wounds in her skin to mine her minerals, felling the great trees for prodigious western appetites, and creating unsustainable biofuel deserts for insatiable automobiles.

Equally disquieting, however, is how much of our experience today is increasingly virtual, controlled by electronic gadgets, computers and television that bring sensory deprivation of touch, smell and even sound. Even our travel is virtual, with no real contact with the land or its inhabitants. There is an exciting world out there beyond the windscreen, if only we can get out of the car and experience it first-hand. How many of us experience listening to the music of a rushing stream, have responded to the haunting owl's call, or have been entranced by the wonder of a Perseid meteor shower in the sky, unblinded by light pollution?

Dwindling water supplies

Before they were seduced by the luxuries of modern living, people the world over followed practices of personal hygiene which were sustainable. In the East in particular, daily collections of 'night soil' ensured the fertility of the land. You can still see this in China and Tibet. In many parts of the world, the 'long drop' privy was the standard toilet. Now, from Aden to Singapore, houses are built with flushing toilets and plumbed showers, causing crises in water supply. One traveller, anxious to impress his Aden family, brought back a smart white porcelain toilet, which stood unused, in the corner of a room.[4]

In May 2008, ocean-going water tankers delivered fresh water to Barcelona, suffering drought from poor water management. Spain's rainfall has markedly fallen, combined with rising temperatures, but local authorities still persist on watering the tourists' golf courses and filling their swimming pools. It is salutary, though, to compare Barcelona's daily water consumption of 114 litres per person with Mexico City's average of 300 litres for its 20 million people.[5] The poorest districts in Mexico City depend on twice weekly tanker deliveries, while the prosperous suburbs are profligate in their water use. Infrastructure problems result in leaks consuming 38% of the supply and there is little recycling or local use of rainwater.

The subject of water is very topical, mainly because usable water is in short supply (see Plate 2). Predictions are now common that wars will be fought over access to water. It is easy to see why. Countries that control the headwaters of important rivers can restrict their flow downstream, like Turkey with Iraq, Israel with Jordan, Syria with Israel, Sudan with Egypt, India with Bangladesh.

Twenty per cent of the world's population do not have clean drinking water; nearly half the world does not have modern sanitation. One hundred cities in northern China now ration water, and Beijing's future as China's capital has been under review because its growth has outstripped its water resources. Even those countries that have sufficient water treat it so badly that, when delivering it to homes, they kill it with chlorine, fluorides and other chemicals, ostensibly to prevent disease. Instead this depresses our immune systems and makes us more open to infection.

How has this come about? Water is in great abundance on this marvellous planet, but only about 0·5% is available as surface fresh water. The rest is salt water, inaccessible groundwater, precious aquifers, or frozen in polar or mountain ice. While the world's population is increasing by 85 million a year, cities are expanding at double the rate due to urbanization. Cities and industries consume the most water (industrial water consumption is set to double by 2025).[6]

Twenty-four countries, mainly in Africa, will not have enough water to meet 2025 projected needs.[7] And, if that is not critical, according to a recent UN report, world population could rise from 6.1 billion in 2000 to at least 8.2 billion by 2050.[8] Today, 1.2 billion drink unclean water, and 2.5 lack proper toilets or sewerage systems.[9] What will be the situation in ten years' time?

Globally, about 70% of water diverted from rivers or drawn irresponsibly from aquifers is used for irrigation. This is hugely wasteful; leaking pipes and channels, evaporation from reservoirs and from irrigation sprays means that about 60% of the water does not reach the plants' roots. China's greatest river, the Yellow River, has run dry and in several years since 1985 has failed to reach the ocean.[10] The once mighty Nile, Ganges and Colorado Rivers barely reach the sea in dry seasons. The introduction of industrial agriculture into India and Northern China has in those areas led to dangerous lowering of the water table.[11]

Water for profit

Perhaps the most telling judgment of our society's ethical standards is our relationship to water. Water has become one of the liveliest traded products on the commodities exchange. To regard water as a commodity is as despicable and immoral as claiming ownership and the patenting of seeds. These are God-given sources of life.

The construction of large dams, whether for hydro-electric power or for irrigation does incalculable environmental damage, as well as annihilating viable human communities. Dams destroy ecosystems and sever the balancing of dynamic energy from one part of the landscape to another. Since 1970, when Egypt's Aswan High Dam came into operation, the number of commercially harvested fish species in the

Nile dropped by two-thirds, and the Mediterranean sardine catch has fallen by 80%.[12]

Vast new networks of supply and disposal pipes must be built in the cities if basic water needs are to be met. Governments, unwilling these days to invest in social infrastructure, are privatizing water utilities, but the results seldom benefit the consumer. A shortage in any essential commodity brings out the profiteers and extortionists. Pro-privatization propaganda reached a climax at the Water Forum meetings in The Hague in March 2000, but the abuses and inadequacies of commercial control have become apparent. Water barons collude to keep prices high, pay enormous bribes to obtain water contracts, and have a low record of efficiency.

One study showed that Swedish municipal water authorities delivered water at around a third of the cost, had operating costs of about half, and produced nearly three times higher return on capital than English private water companies of similar size.[13] After the economic downturn of 2001 several English private water companies experienced financial difficulties and were bought by foreign companies. It makes complete nonsense that essential water supplies should be subject to the ups and down of the financial markets.

In April 2000, the protesting citizens of Cochabamba in Bolivia suffered over 180 casualties at the hands of its own police before their government revoked the right of International Waters of London to impose a 35% price hike in water prices. The government has since reconsidered its policy to privatize all public water supplies.

The great danger to our water comes from the globalization of supply. Multinational companies are unaccountable, and have more interest in profit than in a sustainable environment. A group of water companies tried at the 2001 Hague Water Forum conference to foist a new water order on the world, in effect to encourage water supply to be removed from public control. American companies are negotiating to build dams in India that would displace countless communities and destroy their environments. Three French companies already control more than 70% of the world's private market.[14] Increasing numbers of privatized water schemes are linked to ventures to extract more water through vast dams and reservoirs, with bulk water supply schemes that guarantee profits by requiring consumption regardless of need.

Modern water treatments

Chlorination

Because public water is not treated with the care required to keep water pulsating and alive as Schauberger demonstrated that it must, it degenerates, attracting pathogenic organisms. As a result, the authorities routinely treat it with chlorine to prevent the threat to the community of water-borne diseases. This powerful disinfectant removes all types of bacteria, beneficial and harmful alike, and in doing so, over a long period of time, destroys or seriously weakens many of the immune-enhancing micro-organisms in the body.

It is a major contributor of lowered immune resistance in older people. Medical authorities say that the amount of chlorine is so small that it could not do this, but they fail to take into account that the chlorine accumulates in the fatty tissue of the body, so that the dosage is cumulative, nor that there is a homeopathic action that amplifies the effect on the body. Hot showers and washing machines can produce debilitating chloroform gas which is absorbed by upholstery and carpets. Viktor Schauberger had strong views about compulsory chlorination:

> Those of us who live in cities and are forced year-in and year-out to drink sterilized water should seriously consider the fate of that 'organism' whose naturally-ordained ability to create life has been forcibly removed by chemical compounds. Sterilized and physically-destroyed water not only brings about physical decay, but also gives rise to mental deterioration and hence to the systematic degeneration of humanity and other life-forms.
>
> If we have any common sense remaining, we should refuse to continue to drink water prepared in this way. Otherwise we risk a future as cancer-prone, mentally and physically decrepit, physically and morally inferior individuals.[15]

It is, in fact, not difficult to remove chlorine from our domestic water supplies (see Appendix 1).

Fluoridation

The issue of adding fluorosilicates (fluoride) routinely to drinking water is one of the worst outrages in public health policy. This is not the naturally occurring calcium fluoride that is naturally present in some drinking water, usually at low levels of about 0·1ppm (parts per million). It is a by-product of a number of industrial processes, initially the iron, copper, aluminium and now the phosphate fertilizer industries, and contains also a number of heavy metals; altogether a potent toxic cocktail.[16]

Many dentists claim that fluoride protects teeth from decay. As the disposal of these industrial wastes is very costly, the cynical plan was proposed to add these industrial wastes to public water supplies was started in the USA, lobbied largely by the Mellon family, owners of ALCOA, the largest aluminium manufacturer, and one of the principal fluoride wartime polluters. Starting with Grand Rapids, Michigan in 1945, it was introduced within two years to a hundred cities. Basically a dirty tricks campaign that labelled opposers as crackpots (and during the McCarthy era as left-wing subversives), it has never completed convincing tests, nor produced adequate evidence of its efficacy or safety. 'It was a political, not a scientific health issue' and, like the agenda of the more recent genetically modified foods campaign, became a major US export.[17]

Fluoridation of water supplies is the present policy in much of the English-speaking world, and in a few other countries like Chile, this is permitted, usually at levels of about 1ppm (or 1mg. fluoride per litre of water), but many other countries, where it is often strongly opposed by citizens' groups as well as by scientists, decided that the risks were too high. The five countries that still fluoridate to a large extent are Ireland (75% of the population), Australia (66%), USA, Canada and New Zealand (50%); Britain is lagging at 10% (mostly in the West Midlands and the North-East), but the UK government has now strengthened legislation to enable fluoridation to become mandatory.

The World Health Organization and the American Medical Association were persuaded to back the policy. The FDA (Food and Drug Admin.) has backed off slightly from its 100% endorsement of the practice, due to public exposure of the scam, but today 130 million Americans in 9,600 communities continue to drink fluoridated water.[18]

The addition of fluoride as a policy is justified by the claim that it reduces dental cavities, especially in children. Independent research challenges this claim, showing that the body accumulates levels of fluoride in the bones and certain organs, and there is evidence of increased risk of cancer, brain function impairment, kidney malfunction and premature aging. At higher dosage levels fluorosilicates are an effective rat poison.

Unfortunately fluoride is also added to many processed foods, fruit juice, milk and, especially toothpaste. Fluoride is released into food cooked in Teflon-coated cookware, so the actual intake may be significant, even if you don't live in a fluoridated area. For reasons that are difficult to comprehend, but which are clearly political in nature, many dental and health authorities seem to support this mass medication of whole populations, and politicians seem happy to go along with it.[19]

Barry Groves' study of fluoride policy in the US concludes that:

> Fluoridation is the longest, most expensive and most spectacularly unsuccessful marketing campaign ever to come out of the United States ... This is an example of a cynical alliance between industrial leaders, mainstream science and the politicians we elect, which shows complete contempt for public health and welfare, and indeed for the truth; it shows that we have to be extremely vigilant to preserve our freedoms, and indeed, our health.[20]

Recycling water

We must urgently learn how to recycle water rather than flush it into drains and the sea. The first step is to capture rainwater in water butts to irrigate our gardens. The next is to filter and reuse grey water — that is, domestic waste water. Then comes the acid test of sustainability — to recycle human water waste.

The saying goes 'When needs must' — and Australians, with their dire water shortages, are currently implementing community schemes for using 'grey' water for irrigation. The next stage will be to recycle human urine (after treatment) and faeces also as 'night soil', as has been practised for centuries in Himalayan communities.

Transmuting water's memory

Most communities make genuine efforts to remove physical pollutants from public water supplies, but there are so many organic toxins produced by industrial agriculture, that one is wise to consider good filtration to reduce the dangers of these pollutants and of heavy metals that, sadly, are now common. There are now generally available good and affordable plumbed-in filters that remove most of the physical contaminants. However, what our water treatment policies must urgently take on board is that the physical removal of a pollutant is only part of making water safe.

Typically, in modern cities, public water supplies are recycled as many as twenty times. Even if the physical contaminants have been removed, their vibrational imprint is still carried in the water in its memory bank, no matter how many times it is recycled. Just as water can carry restorative energies, such as in homeopathy, so it can transmit negative or destructive imprints that can cause disharmony or disease in the body.

The purpose of some of the better vortex treatment systems is to recluster the water, in a manner that superimposed natural energies will erase the memory of the water's previous abuse. The vortex, being the enabling gateway between different qualities or levels of energy, allows the water to absorb the etheric energy that surrounds us all. Rather as allowing brilliant sunlight and fresh air to fill a musty room will quickly transmute the stale energy, so the more refined energy always prevails over the coarser. The best domestic plan is a combination of an efficient plumbed-in filter with a vortex-type re-energizing system (see Appendix 1, Water and Health).

Water mains material

Archeological research has shown that in ancient times, from the Babylonians to the Greeks, there was a greater understanding of water and its qualities. In those times, water mains were constructed of high quality wood or of natural stone. As these natural materials became more scarce, the Romans experimented with different metals. Preoccupied with oxidizing corrosion, unfortunately they often used lead, which brought its own problems of lead poisoning, particularly in the wine goblets where the vinegar in the wine dissolved the lead.

Before the expansion of cities during the Industrial Revolution, many water mains in Europe, and even in New York, were constructed of wood, which allowed the water to breathe and to interact with its environment. After the water mains in Vienna were extended to new suburbs between 1920 and 1931, with steel or iron pipes, internally coated with tar, as opposed to the traditional wooden, Viktor Schauberger found that the incidence of cancer more than doubled.[21]

The laminar structure of water quickly disintegrates owing to the chaotic flow through a cylindrical pipe. Friction with the pipe walls heats up the water, decomposing the dissolved trace elements. As the surface of the iron pipes start to rust, oxygen is taken out of the water, and the rust deposits encourage disease-promoting bacteria. The accumulating rust in turn constricts the water flow, so that what is delivered is dead water, which has to be disinfected with chlorine.

The wooden water main

Schauberger knew that water can maintain its vitality and dynamic energy only if it is allowed to tumble about in a spiralling vortical manner. So in 1930 he set about designing a pipe that would actually encourage this movement. It was constructed of wooden staves, like a barrel, which allowed the moisture to seep through, transferring a cooling effect (as in sweating) to the water in the pipe. The spiralling movement was created by a series of guide vanes, which act like rifling in a gun barrel. These were made of silver plated copper to enhance the subtle energies and fluted so as to direct movement towards the centre, thus reducing the heating effects of friction.[22]

Plates 26 and 27 (see colour section) illustrate how this configuration sets up a double-spiral longitudinal vortex, creating a water flow faster than a conventional cylindrical pipe. The centripetal flow of the main water body helps to cool and accelerate it, this heavier water drawing the specifically lighter outer water along in its wake. The centripetal spiralling of the toroidal 'doughnuts' created by the guide vanes makes available oxygen from the main water body, transferring any pathogenic bacteria to the pipe walls where they are eliminated by the aggressive oxygen. The higher quality micro-organisms however, survive, because they require higher levels of oxygen.

It is a brilliant design that imitates the pulsating flow of water in a natural vessel and which delivers water that purifies itself and cools through its motion, eliminating the need for any sterilizing or purifying additives. Ideally, these wooden water mains should be embedded in sand, allowing them to breathe, and protected from light and heat. In such conditions they should outlast a steel pipe.

Water storage

With good water becoming increasingly scarce, it is important to understand how to preserve its quality. Water's enemy is excessive heat and light. Water contains oxygen, a substance that is essential for the processes of growth and decay. Below a temperature of 9°C *(48°F)*, its oxygen is used for growth, above that, to promote decomposition. As the temperature rises above 10°C *(50°F)*, the oxygen becomes increasingly more aggressive, promoting pathogenic bacteria that can give us disease when we drink the water containing them.

A tank that is above ground needs to be well insulated, and painted white to reflect the Sun's heat. If it is mostly below ground, the walls will not require insulation, but the top must be painted white. However, Viktor Schauberger urged us to observe the shapes that Nature uses to propagate and maintain life. Nature abhors squares (cubes), rectangles (water tanks) and circles (cylinders). He said that we should not be surprised that our dependence on these unnatural shapes for storage results in the deterioration of our water. This is probably impractical for larger containers, but we should try a more natural shape for smaller ones.

Because it behaves like a living organism, water needs to be in constant movement to maintain its health. The one container that allows this is the egg-shape. The material of containment is very important because water needs to keep cool; the best materials are natural stone, wood or terracotta. The ancient Greeks understood this, and kept their water (and wine) in *amphorae,* egg-shaped vessels that allowed the liquid to breathe. In amphorae discovered in archaeological digs, grains have been found to be preserved so well after two thousand years that they germinated when planted, proving the effectiveness of the egg-shape for preservation.

Earth healing

There are many traditions that recognize a Spirit of Water. The Saami people of Arctic Scandinavia call her *Mere-Ama*, whose blessing is often sought for good health and fertility. The Water Mother spirit may be identified as a deva who can be sought for advice on personal matters of health and welfare. In many old cultures, working with the Water Spirit was a women's ritual.

Australia has been suffering her worst drought for a century. Alanna Moore, the Australian geomancer, tells of a Buddhist group in 2006 who conducted a rain-making ritual near Bendigo in Victoria:

Monks from the Atisha Centre (the site of the largest stupa in the western world) gathered in bushland beside a dry watercourse for their rain ceremony. It has been one of the driest years on record, yet a few days later a local deluge of some 60 millimetres of rain fell, a friend who attended tells me.[23]

Holy water

Many spiritual traditions celebrate water as a sacred medium. In the desert environment of the Middle East which cradled the great monotheistic religions, water was clearly the precious source of all life. Jesus often referred to the spiritual qualities of water. His miracle of turning water into wine is very symbolic of the key role that water has in any healing activity. The practice of grace before a meal is based on this belief, though many who do it have lost touch with its relevance. Masaru Emoto's experiments showed that praying over a specific sample of water improved its quality. Many of the world's religious practices, going back millennia, are based on this eternal truth.

We shall now look at some of the complexities of our changing times, to see if we can identify the priorities for the near future.

We live by the grace of water.

National Geographic Special Edition, November 1993

17. Water and Climate Change

Where least expected, water breaks forth.
(Dove non si crede, l'acqua rompe.)
Italian proverb

Climate change is symptom, not cause

Human society is in crisis, not from the catastrophic effects of climate change, but because we don't yet acknowledge how we got into this situation. Climate change is not the issue; it is a symptom of our dislocation from the environment, due to our species' hubris which, in the last three hundred years, has grown to mammoth proportions. Nature will always have the last word. Unless we are willing to learn from her, the future prospects for Man are poor. Humility is a rare quality today. There is, as yet, almost no sign of a change of heart; we think we can reduce the effects of global warming through our clever technology. I believe, however, that water can show us how to reconnect with our Source.

Economic and political instability

In 1820 the world population hit one billion. The growth of population since then to over six billion has been possible with the exploitation of fossil fuels. Oil, the most versatile fuel ever discovered, has brought about an unprecedented revolution in methods of transport, and has spawned a raft of technologies from industrial farming to the chemical industry, to computers, medicines and health, hospitals and myriads of consumer goods, world trade and tourism. Our livelihoods are dependent on it.

The creation of enormous wealth has always destabilized human societies. The prodigious wealth that fossil fuels, particularly oil, have produced in the last 190 years has been absorbed by the economically developed nations, with disproportionate excess going into the hands of a few, and in our times has seen the growth of multinational corporate power that is beyond national laws or regulation. Political institutions collude with these centres of power, and the losers are the organic health and wellbeing of human society and the environment.

As demand for oil begins to overtake the decreasing supply, oil production will start to run down (by 2035, it is predicted to be about 60% of the present peak level) and will no longer be a reliable source of energy, for the countries that still have some oil reserves will want to hold on to them, or use them for strategic bargaining. Undoubtedly, wars to control oil supplies will become more numerous. All those countries which have depended on fast economic growth, the USA, China, India and even Britain, as net importers of oil, will be most vulnerable to economic decline. The web of world trade is now so interdependent that the failure of its transport system would be catastrophic.

The population bomb

This is often called 'the elephant in the room' because nobody wants to talk about it. Global population is widely predicted to rise to 9.5 billion well before the end of this century, a level the Earth's resources would have difficulty in supporting, particularly as food production is likely to be seriously disrupted by the effects of climate change.

The availability of cheap oil and gas, coupled with the rise of short-sighted economic policies, has facilitated a rapid and unsustainable expansion of the global economy, which we now see beginning to unravel. These factors will hit all countries, and may dampen down the rate of population growth in the developing countries.

As we have seen, Nature's main priority is balance, and if any species becomes too dominant or grows too fast, Nature will introduce a cut-back by whatever means. Why should Man be exempt from this — just because we feel we are above Nature? There are many means of cutback, from economic collapse, to decimating wars and global pandemics. The main threat to global stability will come from the billion or more people migrating from countries most affected by climate change to the more wealthy societies less affected. The potential for violence will be great. It is likely that, by the end of this century, the capacity of the Earth to support our species may be closer to half the present level of 6.8 billion, rather than the nine billion world population currently predicted.

Water under stress

It would be very hard to overstate the seriousness of the prospects for fresh water all over our planet during the rest of this century. The reports are being phrased in increasingly apocalyptic terms.

Climate change: Fresh water supplies will be endangered by change in the distribution and intensity of rainfall. Global warming will greatly increase the amount of water that the air can hold. This will result in more frequent downpours similar to the monsoons of the tropics in temperate latitudes like the British Isles. Some areas that used to have rain may experience droughts; increased desertification will spread, especially in Africa, Southern Europe and south-west America.

The oceans: The effects of ocean warming are presently poorly understood. One of the trends that give much concern is the effect this has on phytoplankton that are not only the base of the food chain, but also one of the principal absorbers of CO_2, which is stored in their bodies on the ocean floor. They are very sensitive to ocean temperature, preferring a cooler range. The collapse of their populations would affect, not only a wide range of fish stocks, but also the viability of the oceans as a carbon sink.

Deforestation: Forests, especially the tropical, create and recycle rain. The failure of the Amazon rainforest will reduce the rainfall north and south of Amazonia, and the Atlantic shores are likely to lose the moderating climatic influence brought by the Gulf Stream.

The melting of glaciers: The great river systems of India, Pakistan and south-east Asia depend on Himalayan meltwater; there will be acute summer water shortages when the glaciers disappear. Europe's greatest rivers, also, are presently fed by melt from the Alpine glaciers and may well start to run dry in hot summers. Rivers above the 50° latitude are thought to be less affected.

Exhaustion of aquifers: Agriculture in the continental interior of North America has, for the last century, been dependent on the immense stores of ancient water in the deep aquifers below the plains. These have been drained far beyond the point to which natural replenishment can take place. The Australian basin's aquifer is also dangerously low. (See further, 'Regional problems' below.)

Increasing population: Population increase and rise in living standards has put a great strain on water supplies. This is particularly evident in Northern China, parts of India, Indonesia, Mexico and California. Economic immigrants can also strain local resources, and dwindling dependable water supplies could spell famine in many parts of the world, especially in Africa.

Competition for water resources is more likely to bring conflict between nations, even more than that caused by the exhaustion of oil reserves surely, because there are alternatives to oil; for water there are none.

Some nations are in a position to control their neighbours' sources of water (for instance, China with Thailand, Laos and Cambodia, through its control of the Tibetan plateau). This will happen on a local scale all over the world.[1]

The solution to the water crisis lies only partly in learning to conserve supplies and to stop profligacy and abusive practices. We need to regard water as a precious, life-giving substance. The other part of the solution is the need to understand the real significance of water for life; water can show us how we can live sustainably in harmony with the planet.

Water scarcity

The map of water scarcity in regions currently populated shows five areas under severe stress (in rising order): the American South-West and Mexico, the North African coast (especially Algeria), Palestine and the Nile Delta, Pakistan and South India, Northern China. The causes vary from population pressure and economic growth to inappropriate irrigation technologies and change in rainfall patterns. (See Plate 2.)

The American South-West has experienced substantial immigration from Mexico into both southern California and Texas, putting a strain on the infrastructure and on the demand for fresh water. Northern

California has one third of the state's population, but 75% of the state's water resources. Aqueducts and canals, begun in the 1930s, bring surplus water from the north to the farming region of the Central Valley and to the Los Angeles conurbation, which also receives water from the Colorado River, out of state.

Agriculture takes 80% of captured freshwater, leaving domestic, industrial and environmental needs to compete for the remainder. Much of the irrigation effort is wasteful. There have been efforts to recycle 'brown' wastewater for irrigation, but there is resistance to its use for domestic purposes. The main problem here, as in every area of water stress in the world, is the excessive pumping of groundwater, which is non-sustainable, as these resources are irreplaceable.

Mexico City is one of the largest and fastest growing cities in the world. It has exhausted its underground aquifer and the infrastructure has not been set up to bring in adequate supplies of fresh water from the mountains.

Coastal North Africa has seen much higher temperatures in recent years, and a fall in rainfall. Urbanization of Algeria's population has compounded the water problem.

The highly populated Nile Delta is also the main source of Egypt's food, and Cairo has experienced great increase in population. It suffers also from the effect of warming of the whole Mediterranean region. The delta is also threatened by sea level rise.

Israel is a pioneer in soil irrigation directly to the plants' roots. But the water table has dropped seriously throughout the region.

Land under stress

As a result of climate change, the prognosis for the latter part of the century is that land suitable for growing crops will be limited to higher latitudes: while parts of western Russia might still be able to grow crops, areas like Siberia, Canada, parts of Northern Europe and southern South America will become the bread baskets of a much reduced world population. Presumably the taiga pine forests of Siberia would be sacrificed for agriculture. The equatorial rain forests will die out if global warming accelerates.

Apart from drought, the main problem of food security will be our dependence on the monoculture of grain types. Because they are not protected by diversity of species, plant types are vulnerable to global disease, such as caused the Irish potato famine, the problem with wheat rust and other endemic diseases. There is an urgent need for crop breeders to develop a wide range of alternative species, using the wild plant bank.[2] Industrial agriculture has been blind to this predictable situation.

Saline agronomy

It is most likely that, because of the effects of climate change, fresh water supplies in many parts of the world will be become insufficient to grow the food we shall need for future populations. At the moment less than 1% of the world's available water is fresh, most of which is locked in the polar icecaps which are melting without benefit to humanity. Another 1% is brackish (less salty than the oceans), and then there are the unlimited oceans. Many plants can be grown in saline water, for food, for minerals and for energy.[3]

Water in the landscape

Water creates vibrant life appropriate to each climatic landscape. Whether this is driven by the intelligence of Nature or by Cosmic 'plan', or even from some inherent intelligence of water itself may be beside the point. Man has irreparably damaged water's natural role to optimize the fertility and biodiversity of the landscape.

When we allow water to 'do its thing', as Peter Andrews did in Australia (see Chapter 18), apparent miracles can happen. We have become blind to the fact that water is an integral and essential part of the landscape. We often find natural water features inconvenient, messy, or accompanied by wildlife we don't appreciate. So the mangrove swamps defending the tropical coasts of Africa and South-east Asia were removed to make way for mono-cultured shrimp farms and what happened? Those coastline communities were devastated by the tsunami of 2004.

The most prolific natural environment is the equatorial tropical forest. Temperate climates cannot sustain such high levels of biological wealth, but temperate latitudes are more sought after for human settlement and agriculture.

People have always chosen rivers as their main focus for settlement, as they provide fresh water and easier communications. This is fine as long as rivers are not badly regulated or polluted, and when we don't extract too much of the fresh water. Population density has increased to the point where water features are crowded out of the landscape. Apart from increasing biodiversity and ecological richness, water features act as a sponge to absorb high levels of precipitation.

Rising temperatures increase evaporation from the oceans, resulting in higher atmospheric humidity which will produce more torrential and unpredictable rainfall. Where will this go if we have removed the sponges which allow slow release of water? Monoculture does not make a good sponge. Farmers who plant winter wheat on exposed higher ground have found their soil suffers from damaging erosion. The removal of forest or woodland on watersheds is an irresponsible form of Earth vandalism, exceeded only by removal of the fragile rainforests.

Swamps, marshes, ponds, bogs and wetland all encourage the richest diversity of fauna and flora, starting with the tiniest forms of life, the base of the food chain — bacteria, bugs, insects and microscopic invertebrates. They have been disappearing in Europe with the demand for commercial use of the land. These need to be brought back as part of a system of water purification and recycling. So septic tanks and human effluent can be treated by horizontal or vertical flow reed beds, whose plants take oxygen down to the roots to aerate the water. Toxic farm effluent can be treated by serpentine channels of ditches planted with willow. Finally, a staircase of flowforms oxygenates and energizes the cleaned water.[4]

Permaculture, an organic form of cultivation with minimal inputs, optimizes biodiversity, with a variety of perennial plants and the recycling of water. This system makes a good sponge, and often uses water features.

We have all heard that the Earth is warming at an alarming rate, unprecedented for many thousands of years. The increase in CO_2 emissions is the main contributor, but this is exacerbated by secondary negative feedback mechanisms that act like a row of dominoes pushing each other over, such as: warming in the Arctic tundra melts the permafrost which releases vast stores of methane (the most powerful greenhouse gas) that have been safely locked up for millennia, which accelerates global warming, etc.

The changes that will affect humanity most are in rainfall patterns. As we have seen, rapidly increasing demand from population growth and urbanization has in any case created shortages of fresh water. The changing patterns of precipitation take us by surprise. A rise in global temperature will bring an increase in global rainfall as the atmosphere can hold a higher amount of water vapour. These downpours are not uniform. They cause intense rain in some places, but not others, often unpredictably.

Very often a geographical shift may be caused by a change in the behaviour of the ocean currents, the North Atlantic jet stream, or the Bay of Bengal monsoon patterns. Climate scientists have warned that a maximum of the El Niño effect in the 2010s will exacerbate global warming, bringing much higher temperatures.

The atmosphere and the oceans together behave as a single organic whole, changes in one part of which can affect the opposite side of the globe. So a hotter tropical climate can result in increased desertification in the centre of continents (for instance, central Asia), as well as bringing more extreme precipitation; and the destruction and dieback of the tropical rainforests will result in more extreme atmospheric conditions, producing more violent hurricanes, typhoons, deluges and floods.

Regional problems

Australia

The early colonists destroyed the land's water balance by draining wetlands to create pasture and cropland. Soil fertility was badly affected and further damaged by artificial fertilizers. Victoria's rivers were systematically cleared out which affected the biological balance of the region, encouraging invasive species. The Murray/Darling has been overwhelmed by carp.[5] Logs and other materials create turbulence and deep holes which are important for fish to thrive.

The Australian subcontinent has been experiencing a warming trend for some 30–40 years. The last eleven years have brought severe drought conditions to the south and parts of the east coast. Significantly, the per capita consumption of water is the highest in the world, yet Australia is also the driest continent.

A campaign for stringent water conservation and rainwater harvesting for use in cooking, washing and cleaning has been implemented in the cities. When properly filtered, this water is also used for drinking. The national norm is a three minute maximum for showering, and no car washing. Many people have installed septic tank systems with aerobic digesters and filters that recycle toilet water many times over.

Some authorities are experimenting with solar power desalination plants to convert seawater into drinking water. Australian climatologists believe that drought conditions will persist and worsen. Ross Young, executive director of the Water Services Association, writing in *The Australian,* when asked what can be learned from the Australian experience, said: 'When climate change begins to impact water supplies, it does so in a far more rapid and dramatic manner than any of the experts ever predicted. That's why everyone must be proactive.'

El Niño (dry) and La Niña (wet) are oscillating patterns of atmospheric pressure and warm ocean circulation that affect Australia's rainfall on its north and east coats, normally with years of alternating storms and drought. In the last forty years there has been a dramatic decline in winter rainfall in south-eastern Australia, which many believe is caused by a breakdown in the normal El Niño cycle. The Murray-Darling basin, the nation's breadbasket, which accounts for 41% of its food production, has suffered a severe drought. The basin covers an area of over one million km2, with a catchment from Queensland's tropical north to the Darling River, and from the Murray's source in the Snowy Mountains in the east down to South Australia near Adelaide.

In 1949 the federal government created the Snowy hydro scheme to generate hydro-electric power and capture water from the spring snow melt in two large lakes in order to regulate the flow of the Murray River, so that farmers downstream could draw water for irrigation in their dry summers. For a time this worked well, but the allocations were not scaled down as the drought began to bite, and a combination of over-extraction and decreasing winter rainfall has reduced the Murray to a trickle that often does not reach the ocean.

Australia's experience of such a prolonged dry spell may be a foretaste of what is to come. Tim Flannery, the Australian climate change guru, believes that crippling drought could now happen in parts of the world like Northern India, Northern China or Western America, precipitating water crises all over the world.

Asia

Pakistan's Indus Valley has been dry for centuries, and with increasing industrialization, the underground sources are becoming seriously over-exploited. In both Pakistan and even more in southern India, the introduction of industrialized agriculture, and water hungry industries (like Coca-Cola) have increased the water stress in a region where rainfall is becoming more unpredictable.

Climate change is particularly worrying in Northern India. Across the country, from Gujarat to Hyderabad and in Andhra Pradesh, 'the rice bowl of India', the monsoon season figures fell 43%.[6] There are reports of communities at war, and of some drawing water being 'hacked to death by angry neighbours accusing them of stealing water'. In Bhopal, where 100,000 people depend on water tankers, fights break out regularly.

Northern China suffers from the most serious water stress situation, especially around Beijing, which has seen enormous population growth and industrialization. The water table in the region has shrunk alarmingly, about a foot a year.

An ambitious plan to bring water from the moist Yangtse basin in the south up to the dry lands above the Yellow River was envisioned by Mao Zedong in 1962. It was to involve building three channels, each more than 600 miles long. They would be twice the cost of the Three Gorges dam, and three times the length of the railway to Tibet. The project has been delayed because of ecological, political and financial pressures. The pollution of the Yellow River is very serious. Regional political authorities feel their needs are greater than those of the northern cities. It may well founder because of the controversial problems.[7]

China's great push towards record economic growth could well become stalled for want of sufficient water for its people and its industry. Its great rivers are drying up. The famous Yellow River much of the year does not reach the ocean. China will also suffer, like

Pakistan and Northern India, from the ever decreasing melting of the Himalayan snows and glaciers, the greatest store of fresh water on that continent.

Europe

The Mediterranean countries will become hotter and even drier. Spain particularly has been feeling the change already. Spaniards are profligate in their use of water to support tourism. They have built hundreds of golf courses and thousands of swimming pools. In 2007 the country suffered its worst drought for sixty years. Catalonia, of which Barcelona is the capital, has been the worst hit, with its reservoirs almost empty. In May 2005 this city chartered six huge tankers to bring fresh water to alleviate the shortage. Tourism may well slump as oil shortages bite into the cheap airlines' business, but the Spaniards will have to become better water misers.[8]

Greece had frightening fires around Athens in 2009. The danger is that the damaged land could degrade into desert unless it is quickly reforested. Tree planting is always the best protection against climate change in warm climates.

Britain is predicted to have wetter winters and hotter summers. It will certainly be subjected to more violent weather, with floods, storm surges, coastal erosion and flooding. If the Gulf Stream slows down much more than it already has, the British Isles may lose some of the warmth it has taken for granted for several millennia. The position of these islands at the boundaries of several climatic systems makes the changes they face more unpredictable.

North America

Most climatologists agree that the interior of the United States will receive less rain, creating serious problems for food production. Southern California and the South-west, and the states west of the Rockies are already experiencing droughts. There is likely to be an increase in the number and severity of typhoons and hurricanes.

In 2009 the first hearing of the world's top scientists to the US Congress warned of severe droughts throughout the West, searing heat in the cities, dropping water levels on the Great Lakes and increasing heat-related health problems. Big battles are foreseen between the Democratic leadership alerted by concerned scientists and climate-change denying politicians.[9]

One consequence of these droughts, coupled with rise of population and poor land management by farmers and oil and gas companies, is an increase in dust storms. When dust settles on the mountain snow packs which normally reflect the sun's heat back into space, premature melting of the snow takes place which changes the blossoming and growing times of vegetation, affecting Colorado farmers.[10]

Many coastal cities, like New York, Los Angeles, Seattle, Portland, will be at risk from rising sea levels. Much of Florida is low lying and will gradually be flooded by a rising sea level. In addition, the coast between New York and South Carolina is falling at about 15cm a century, due to isostatic adjustment after the last ice age, which means greater loss of coastlands here.[11]

Reduction in the polar icecaps

How could the optimum climate be designed to encourage maximum biodiversity (the evolutionary imperative) and one that favoured the emergence of *Homo sapiens*? And while you're at it, one that is self-correcting in the event of variation in the Sun's radiation? By creating polar icecaps!

The polar ice sheets have existed for the last ten million years, and have coincided with the development of unprecedented biodiversity, the growth of tropical and temperate forests, with great fertility and soil depth in the temperate latitudes. The balance between hot and humid tropical latitudes and the polar ice has allowed temperate climates to develop in the middle latitudes. This period also has seen the emergence of the higher mammals and hominids. Will they have a future if the icecaps disappear?

The climate of the past few thousand years has been very kind to humanity, allowing us to spread to almost every corner of the globe. With the availability of fossil fuels, we have enjoyed the optimum conditions for supporting 6.8 billion people. This will all soon change. A shift in global temperatures reduces the range of environments that can sustain human life. During the last ice age when the difference in the average world temperature between then and now was 4C°, humans were restricted to lower latitudes. Global warming will have the opposite effect, driving people to higher latitudes.

The initial cause of climate change is generally acknowledged to be an increase in greenhouse emissions. When the average increase in global temperature exceeds 3C°, the remains of the equatorial rainforests are likely to die back, completing the process we have initiated through mindless deforestation. The forests' destruction may turn out to be a more potent cause of world climatic change than CO_2 emissions, for they influence the world climate more than we realize, and will make the equatorial climate more extreme, and all climates less predictable.

The warming effect of climate change has been most marked in the Arctic, where it has been three times the rate of the Earth as a whole.[12] The sea ice has been shrinking in the summer months at an unprecedented rate.[13] The Northwest Passage linking the North Atlantic to the Bering Strait and onwards by the Northeast Passage to Europe is predicted to be navigable within a few years.[14]

The sea ice-shelf reflects most of the Sun's radiation (the albedo effect). Its melting allows the Arctic Ocean to absorb the Sun's heat, a positive feedback effect that cumulatively, year by year, accelerates the rate of warming in the Arctic and releases methane from the sea bottom.[15] Greenland's glaciers and ice cap are melting at an ever-increasing rate, causing enormous pools of fresh water which could close down the Gulf Stream if they escaped out of the Arctic basin, south of the Spitzbergen ridge. The melting of all of Greenland's ice would raise world sea levels by over seven metres, inundating most coastal cities and fertile agricultural land. The Antarctic ice cap is melting at the edges, its rate currently predicted to be much slower than that of Greenland's ice.

The IPCC (International Panel on Climate Change) which revises its predictions regularly, now looks towards an earlier dramatic rise in sea level. Their reports look optimistic because they are usually already out of date when published. Warming of the ocean also increases its volume. The IPCC estimates that, by mid-century, significant flooding of centres of population would have started. Minute changes in ocean temperatures have a dramatic effect.[16]

Portent of the sixth mass extinction?

We understand very little about the oceans, which contain 90% of Earth's biomass. Barely 1% of their life forms have been identified and studied. Only recently we have learned that they absorb more CO_2 than land vegetation, especially in the colder waters. The oceans' ability to absorb CO_2 will be compromised by temperature rise. Alarm bells are now sounding among the world's oceanographers at new research indicating that the Arctic Ocean is becoming more acidic as the CO_2 discharged from the water is turning into carbonic acid, posing a threat to the important base of the oceanic food chain like the phytoplankton, which thrive in cold water.[17]

This could lead, later this century, to a collapse of the rich biodiversity of the oceans, which would affect biodiversity on the land (as part of the same biosphere) already suffering under unprecedented species' loss. The implications of species' loss are apocalyptic.

The Great Barrier Reef, one of the Seven Wonders of the Natural World, has one of the greatest densities of biodiversity, including many endangered species. It has already suffered bleaching from increased sea temperatures, and a number of scientists have predicted its demise as a living habitat as soon as 2020.

Man's impact on the Earth

Some geologists are now calling the last 12,000 years the Anthropocene period because in these years Man has completely changed the face of the Earth, is altering its climates and is now precipitating a major collapse of species. This began with the tilling of the land. Wholesale deforestation accompanied the development of agriculture, which caused a release of CO_2 from plants and soil into the atmosphere. Prof. Iain Stewart believes that this gradual increase in greenhouse gases has maintained the momentum of the present interglacial period, which otherwise would have probably ended by now.

Agriculture brings with it significant loss of topsoil and reduction in fertility, as well as, in modern times, widespread pollution of the water table. It is insatiable in its demand for water. The great natural water cycles have been seriously disrupted by Man. There are very few major rivers that maintain their natural energy levels, flow any more with complete freedom — or even reach the ocean. The exceptions are in the remote tundra areas of Northern Canada and Siberia.

The last 12,000 years, with the tectonic uplift of the Himalayas and other more recent Asian mountain systems (augmented by the still orogenically active older Rockies and Andes), have also seen a great increase in the amount of weathered minerals deposited on the land and in the oceans. Iain Stewart has suggested in a recent television series 'How Earth made us' that these deposits, by enhancing carbon sequestration must, until now, have helped keep the planet cool and fit for life.

Water as a source of energy

Wind power has claimed the greater part of investment in renewable energy. Its main disadvantage is the intermittency of effective wind. On the other hand, water provides a much more reliable source. Small scale hydro-electric plants were common on private estates a century ago, and larger systems came in the 1930s. The mammoth projects have had a damaging effect on the environment, whether the Hoover Dam on the Colorado of the 1930s, or the Three Gorges Hydro-electric project on the Yellow River in China that we mentioned in Chapter 6.

The development of wave and tidal power projects have been slower because of the research in the siting of projects, and the higher development cost of the machines. But there is no reason why the costs will not come down, for sites such as the Pentland Firth off the Orkney Isles, the Severn Estuary in the West of England or the Bay of Fundy off Nova Scotia, all of which have tremendous potential for energy generation. We shall need all the renewable potential we can find to replace fossil fuel sources.

Viktor Schauberger developed implosion machines which produced power multiples in excess of conventional generators. They were not very stable, and no-one has been able as yet to replicate them. Both Viktor and his son Walter said the problems people ran into stemmed their trying to build these appliances without the humility to study and learn first from Nature.

Some have developed simple adaptations of automobile engines to run on hybrid fuels — water and gasoline (petrol), or even on water alone (through hydrogen generation or 'Brown's gas'). A number are described on the Internet. I believe that when the time is right and we are ready, Nature will release some of her secrets to us for power generation; but we're not there yet![18] There is even intriguing research that salt water can be ignited when exposed to a radio frequency beam![19]

18. The Future of Food Production

Organic soils have greater capacity to retain water as well as nutrients such as nitrogen. They contain significantly more nutrients — vitamin C, iron, magnesium and phosphorus — and significantly less nitrates than conventionally tilled soils. Organic soils are also more efficient carbon sinks, and organic management saves on fossil fuels.

Mae-Wan Ho, *Food for the Future*

The availability of water for food growing will be one of our greatest challenges. We have seen how water works in the natural environment and in organisms. Above all, we have seen how important are the water cycles. Nature is the great conservator, the recycler, never wasting energy. One might even suggest that man invented the Second Law of Thermodynamics in order to justify his prodigal, unthinking ways!

We shall end this last part of the book by meeting head on the real issue of what we do with our water, for at least two-thirds of our precious fresh water goes to food production. The saga of how we use water for agriculture illustrates as much as in any area of human activity how our present way of thinking and dependence on unsustainable practices will be our own undoing.

World food trade distorts water resources

The water consumed in the production of food and other raw materials is called virtual water. As an illustration, one pound of beef requires 1,857 US gallons of water; a pound of sausages 1,392 gallons; a pair of blue jeans 2,900. North America is the biggest exporter of virtual water, followed by South America and Africa. Asia and Europe are net importers. North America, which exports 28 trillion US gallons of virtual water a year, mostly to Asia (in the form of soybeans, wheat, beef and coffee), will find this will become unsustainable as domestic water shortages start to bite. Though Europe is a net importer, the region trades 60 trillion US gallons between its nations. *(National Geographic,* April 2010.) A net exporter, like Spain, which is suffering from increased water stress, exports water it needs for itself with its food exports. The water resources of a country like Britain, a net importer of food, will suffer more water shortage when it is forced to grow more of its own food as rising oil costs make long distance shipping too costly.

The practice of irrigation

The Earth has accessible water storage systems in its lakes and groundwater. Most of its subterranean water is too deep to exploit, but man has for millennia bored into the Earth's surface to obtain water. When food growing methods were less demanding it was a sustainable practice, borrowing in lean times, and being topped up in times of plenty. However, the great continental aquifers (which are mostly at least a million years old), are now being dangerously depleted, especially in North America, China and Australia.

Between 60–70% of fresh water used is taken by agriculture, whose demands in many parts of the world outstrip sustainable supply. Irrigation from rivers is the main source. With the increasing demand from industry, many of the world's largest rivers now do not, for much of the year, reach the ocean.

Irrigation, for the most part, is very wasteful, most of the water failing to reach the plants' roots, but evaporating en route. We don't know how to cherish water, regarding it, as we do most things, as a resource which is ours for the taking, whenever and however we want. Like air, it should be shared!

Meanwhile our fresh water use is increasing at twice the rate of population increase in many parts of the world. Climate change will redistribute rainfall and over-consumption is threatening the world's natural storage systems. It is hard to see how the required revolution in our awareness can take place while the population time bomb is ticking away.

The likelihood of water wars has become a frequent topic of debate.

Until humanity has learned that negotiation is more effective than aggression, neighbouring countries will compete to control the waters of a shared river. The increasing scarcity of water will trigger conflict greater than that for oil. The most likely are the rivers of the Middle East and parts of Asia and Africa.

Agriculture in the valleys of the great Asian rivers, including China's, has depended on the annual melt from the Himalayan glaciers. These are now receding, which will put at risk the reliability of future water supplies. Increasing desertification from climate change is now being predicted in the USA and parts of Asia.

The next generation will see the biggest revolution in agriculture and food production that the world has experienced for a century. The rise of industrial farming techniques of the last 100 years has been totally dependent on cheap and easily available oil. World production of oil has now peaked and demand is starting to overtake supply. Future supplies will be of poorer quality, at escalating cost, and subject to tricky geopolitical bartering, as the cooperation of those states that now have reserves cannot be counted on. They may wish to retain their remaining oil reserves for their own long-term needs. (The UK and the US in particular, in fomenting war in the Middle East and Afghanistan, and their unconditional support for Israeli expansion, may find they are denied Middle Eastern sources of oil.)

Modern farming is highly mechanized, with high consumption of fuel, and using complex machinery that demands large amounts of fossil fuels to manufacture. Even more desperately, modern agriculture depends on oil-based chemicals — fertilizers, pesticides, insecticides and herbicides. Like the commercial airlines, they will soon find themselves unable to keep their whole system going.

An additional danger is the rush of big business to produce genetically modified crops and transgenic foods, using inadequately researched technologies, backed by opportunistic politicians and by reductionist scientists who refuse to heed the Pandora's box dangers involved.[1]

The damage that industrial farming has done to the world environment is incalculable, causing massive soil erosion, loss of biodiversity, fertility and rainfall; desertification, pollution of the water table, species' loss; mudslides and flooding; loss of woodland and forest. There is so much power vested in the industry that it that a voluntary change to sustainable practice is unlikely.

If you can't see the consequences of your actions, if all your life you've held a myopic worldview, a change to the long view would be like St Paul's conversion on the road to Damascus — and probably unlikely to happen. Our profligate exploitation of fossil fuels is reaping its own rewards, and most economic activity will have to localize, including food production, and revert to more labour-intensive patterns. We shall examine how a more enlightened practice of agriculture could solve our food problems.

Reading the landscape

If only we could see that water is the essential life blood of the planet and cannot be separated from the natural environment, we could then start to work with Nature whose husbandry of water is so efficient.

To be effective, water management has to take into account the whole eco-system. Above all, there must be an awareness of the natural landscape, which over the years develops a balance between woodland, wetlands, naturally flowing streams, lakes and natural drainage. What modern farming at all scales tends to do is heedlessly impose a totally foreign and highly manipulative imprint on the landscape, whose health is consequently bound to deteriorate.

On the 'planet of water', it comes as no surprise that the land skin is water-dependent. Its absence causes desertification. It's not always clear how the important water cycles are disrupted by human activities. The health and fertility of an eco-system is usually dependent on the relationship between the river and the landscape through which it flows.

Australia has one of the worst records of land exploitation in modern history.[2] The state of much of the land the first European explorers described in the southern part of the country before 1850 was of verdant forest and sweet rivers. Deforestation and extractive agricultural practice sand over-stocking have transformed much of the sub-continent into an arid, infertile land. It is probable that the twelve-year drought that the wheat-producing region of the Murray-Darling basin in south-eastern Australia, has been suffering is would not have been so bad without that history.

The early European settlers in Australia didn't like standing water.

They drained the upland ponds and marshes, created irrigation canals and speeded up the stream flow in this once-fertile area, not realizing that these manipulative practices would destroy the fertility and biodiversity, reduce balanced rainfall, cause eroded gullies and excessive salinity, and isolate the rivers from the landscape.

Peter Andrews was brought up near Broken Hill and learned from aboriginal elders how to read a landscape to learn where water wanted to flow. He is a breeder of thoroughbred horses. As soon as he acquired a 20,000 acre grazing property in the Upper Hunter Valley, New South Wales, thirty-five years ago, he knew what he had to do to revive the seriously degraded landscape. It took him several years to see where the water would have flowed when the land was healthy. He led the water back to those places where it used to be and reconnected the streams to the flood plain so that they could bring the minerals and the energy to nourish plant life.

This was not an expensive operation; it was sustainable and required no chemicals or costly engineering. The secret was to restore the natural pathways of water in the valley; to slow them down and allow them to recreate ponds, marshes and the riparian vegetation — all the elements that produce a healthy and biodiverse sponge that would restore fertility to the floodplain and slowly release water in times of drought, keeping the underground salts at bay.

The result is that, in today's severe drought in the Murray-Darling basin, Andrews' land is a green oasis surrounded by aridity. Political bureaucracies and agricultural establishments have difficulty understanding the idea of wholeness of the land. Having fought official scepticism for a generation, his patience is being rewarded with acclaim from many quarters which sees his Natural Sequence Farming methods as the answer for Australian agriculture facing the challenge of climate change.

Peter Andrews' advice is that you can't beat Nature's way; that we have to learn how to study the individual landscape to understand how and where its water wants to flow, as this is the requirement for ecological balance.[3]

It is nothing short of miraculous that water seems to have an in-built knowledge of what functions to perform in each environment that it moulds. But water is also the servant of Nature. This becomes apparent when you allow Nature to reclaim land that has been manipulated by

Man. In the extreme case of chemically fertilized industrial farming it becomes a desert where no birds sing, nor butterflies flutter.[4] When the rainfall becomes adequate, the reclaimed land soon becomes vibrant with life.

Water is sick (*vide* Attenborough) because we prevent it doing what it needs to do to promote and nourish life. What is desperately needed is a new profession of 'water wizardry' to have committed people develop a combination of the skills of the Aborigine shaman, the insights and understanding of quantum biology and the vision and passion of Viktor Schauberger.

Nature is now calling on Man to take up his Earth stewardship through healing the land. This can best be accomplished, as Peter Andrews found in Australia, by encouraging water to bring the healing. His example should be studied by all agronomists and policy makers.

On 'the planet of water', Nature has always ensured that there is sufficient fresh water to nourish all of life. Much of this needs to be stored in wetlands, marshes, lakes and glaciers to be available during dry periods. Healthy land is absorbent. One solution to the coming water shortages in the unpredictable future is to restore the essential sponge nature of the land.

Biological agriculture

The health of the topsoil is the most important factor in sustainable agriculture. Topsoil is created by decayed vegetable matter, and can vary in depth from a few centimetres to several metres. Forests created the deep soils of the world over millennia, but many of these have shrunk by as much as 80% in the last two hundred years through our disastrous agricultural practices, or in extreme cases the whole soil profile has been literally washed or blown away.

Under natural conditions the friable soil is populated with a multitude of microbial and invertebrate fauna and is usually capped with a layer of humus, formed of decomposing leaves and other vegetable matter which retains water like a sponge and is a valuable protection against drought. This rich mixture of life-forms makes up a processing factory essential to soil health and fertility, and everything should be done to help it flourish. We can never put enough organic material on the soil.

Turning over or digging into the topsoil disturbs this rich biological treasure chest. The use of powerful modern tractors has turned the ancient art of ploughing into land abuse. In the more purist forms of biological agriculture, a no-dig practice is being rediscovered.

Intensive farming practices have led to the exhaustion of the soil's minerals and they now rely on the use of artificial fertilizers. These destroy the rich biodiversity upon which the soil's fertility and the vibrancy of its energy depend. The land becomes virtual diversity deserts, without earthworms, bees, birds, insects or wildflowers.

Industrialized farming (at least ninety per cent of total agriculture in most countries — over 97% in the UK) is dependent on the chemical fertilizers nitrogen, phosphorus and potassium (NPK) that have escalated in cost with the rise in the price of oil and which are becoming uneconomic to use. It is also highly dependent on oil as a fuel for machinery and transport, and has a high consumption of electricity. The rise in costs through Peak Oil (when demand overtakes supply) is already beginning to hit conventional farming.

As for energy, it is becoming doubtful if we can develop sufficient alternative supplies to replace the vanishing oil and natural gas. So we need to develop new farming methods urgently.

Governments, advised as they are by and large corporations and their compliant scientists, seem to be blind to the approaching crisis to our energy and food supplies. The food supply is 80% controlled by the supermarkets which operate on a 'just-in-time' basis, and long supply lines that require heavy fuel consumption.

Are we prepared with an alternative? Growers with any foresight should be converting, as a matter of urgency, to organic, low input methods. Industrial farming is water intensive with its wasteful irrigation methods. Organic methods of cultivation favour much smaller scale units which can supply local demand and are essentially more energy- and water-efficient and generally more productive.[5]

Soil remineralization

Agriculture was developed in the Middle East's Fertile Crescent, which benefited from the annual floods bringing minerals down from the eroded mountains. The early colonizers of the New World had the advantage of mineral-rich soils that had never been cultivated. In Europe, however, where cultivation had proceeded for centuries, the soils were becoming depleted of their minerals. This was accentuated by the intensive farming practices introduced in the mid-twentieth century.

In 1894 an agricultural chemist, Julius Hensel, published *Bread from Stone,* which described the beneficial effects of fertilizing with rock dust, a by-product of highway building. His book was bought up and destroyed by the new chemical fertilizer companies who felt their profitable business under threat.

Ideally ground in a cold process that retains its inherent energies, this rock dust is composed of finely ground, mainly igneous rocks, like granite and basalt. Because of their broad range of minerals, trace elements and salts, when spread on the ground, the dust or granules encourage a wealth of different micro-organisms.

Remineralization with rock dust was practised in Switzerland 160 years ago. Its reintroduction was been encouraged by the appearance of John Hamaker's and Don Weaver's book *The Survival of Civilization,* in 1975. They describe the importance of mineral and trace elements to plant growth and quality. John Hamaker tells how he was able, through their use, to increase the depth of the topsoil at his Michigan home, from about 10 cm *(4in)* to about 1.2 m *(4ft)* over a period of ten years. My wife has found that sprinkling rock dust on her beds has substantially increased the quality and size of her vegetables. Rock dust is normally available as a by-product of road metal quarries.

Rock dust has paramagnetic qualities — a weak attraction to a magnet which increases the soil's ability to attract and hold energy. Water has a diamagnetic charge, a weak magnetic charge that repulses a magnet. The dynamic interplay between paramagnetic and diamagnetic (*yang* and *yin*) energies introduces productive oscillating energy effects in the soil.[6]

Organic farming

Organic farming normally uses manure, farmyard slurry and composted vegetable matter to increase the soil's fertility.[7] The introduction of chemical fertilizers in the nineteenth century soon supplanted the traditional organic methods because it was much less labour intensive, and appeared to give higher crop yields. A few farmers retained the traditional methods and, as the evidence has built up of the pollution of the water table and rivers by the chemicals, there has been a renaissance of organic farming in the last fifty years.

The sustainability of organic farming derives from the recycling of organic material to maintain its fertility, just as in a natural forest. Modern organic composting tends to use green vegetable matter rather than dried, interleaved with layers of earth. Significant heat is generated in the pile in this way. Viktor Schauberger believed that as the heat discourages earthworms, it will not be of the best quality. His preference is a cold process that produces a higher content of protein and immaterial, what he called 'fructigenic', energies. He also believed it important to protect the compost from element-hungry juvenile rainwater that will tend to leach out some of the nutrients.

Organic farming is more labour intensive than conventional. It requires greater strength than driving a tractor. A shift to organic would demand substantial numbers of younger people to become growers (in Britain the average age of farmers is over sixty). Organic agriculture and localized food systems mitigate 30 percent of the world's greenhouse gas emissions and save one-sixth of present energy consumption.[8]

It is such a threat to an agricultural industry worth trillions, and to the trillions invested in genetic modification, that their spreading of misinformation and lies about organic farming and its ability to feed the world safely and sustainably, is delaying the urgent need to change.

Permaculture

This modern organic movement started in the mid 1970s, by Australians Bill Mollison and David Holmgren in the belief that if people were to feed themselves sustainably, they would need to reduce reliance on industrialized agriculture. Where industrial farms use technology powered by fossil fuels, and each farm specializes in producing high yields of a single annual crop, permaculture stresses the value of low inputs and a rich mixture of perennial crops. The method is particularly suited to smallholdings.

More usually, a permaculture garden is for a family or a community.

Ideally, it is designed in the form of a forest garden, in which the trees are spaced out to allow the growing vegetables or fruit bushes in between. As near as possible, the design is predicated by what Nature prefers, above all stressing biodiversity. Worms, instead of the plough, turn over the soil. As in a forest, Nature provides the fertilizers free. The rich biological diversity of permaculture gardens tends to attract a rich bird and insect life, and to produce abundant crops.

A permaculture garden may look messy and disorganized, but it is planned with great forethought; the better the plan, the less upkeep is required. Typically hand tools rather than spades are used. Every plant is important in some way: some to provide essential nutrients like potassium, nitrogen or phosphorus; others control pests, attract beneficial insects or retain moisture. Plants may be grown in tiers to increase the density of food production.

The productive garden is one with the greatest biodiversity — teeming with life and health. An acre of garden designed for maximum production can feed ten people — double the output of a conventional farm. Grains may be less easy to grow, but a four-acre nut wood can produce two tons an acre, as much as four acres of organic wheat.[9]

The Transition Network model for meeting the crisis of oil depletion encourages an abundance of small scale home and market gardens for food production, which challenges the hidden costs of transporting food over large distances to their place of consumption and maximizes localized production of food.

Recently, permaculture is becoming a significant subject in the university curriculum. It covers all the issues of community and the sustainable society as well as food growing, and was the impetus for starting the Transition Towns movement.[10]

New models

Mae-Wan Ho has proposed the idea that all sustainable systems act like organisms. They are sustainable precisely because the energy they produce moves round in a cycle and is kept within the system, the surplus of one part being used by another, so there is very little waste.

In her Dream Farm 2 an integrated food and energy systems operate on the same organic circular economy (see Figure 22). The central component is a biogas digester to process livestock manure and waste water. Biogas is 60% methane that is used for cooking, heating, generating electricity and driving cars. The solid residue is a rich fertilizer for crops and mushrooms.

The waste water from the digester goes through an oxidizing cleanser with algae which produce oxygen through photosynthesis to oxidize the remaining pollutants. The cleansed water goes into the fish ponds; this nourishing water is used to irrigate crops whose wastes go to raise earthworms or compost, or fed back into the digester. Algae from the pond can be harvested to feed the hens, geese and ducks. The remains of the mushroom harvest are fed to livestock, with the crops, and their manure goes back to the biogas digester to complete the cycle.

The methane production can be augmented with other renewable sources like solar panels, windmill and micro hydro-electric. Such a farm could feed a thousand people, and be totally energy self-sufficient. If Dream Farm 2 were universally adopted around the world, Dr Ho claims that it would have the potential to mitigate 56.6% of greenhouse emissions and 50.5% of energy use.

Biodiversity

It is the principle with which Nature works. Biodiversity can also be appreciated on an energy level. The greater the biodiversity and the complexity of interconnections between classes of organisms, the more efficiently is energy exchanged within a system and the higher the quantum effect. The concept of biodiversity is often as poorly understood as that of sustainability, because they are part of a holistic worldview that is hard for us to understand today with our disjointed educational system.

The tropical rainforests are such a powerful energy exchange system because of the extraordinarily rich biodiversity of its fauna and flora. When we degrade those precious forests, the whole Earth system suffers and water supplies are put at risk, as we are now being shown. The natural temperate forests also thrive on biodiversity, but at a lower energy level, as the Sun's power is less at those latitudes.

Nature is so successful because she operates on closed energy systems, with no waste; everything is recycled. If this were not so, the Earth

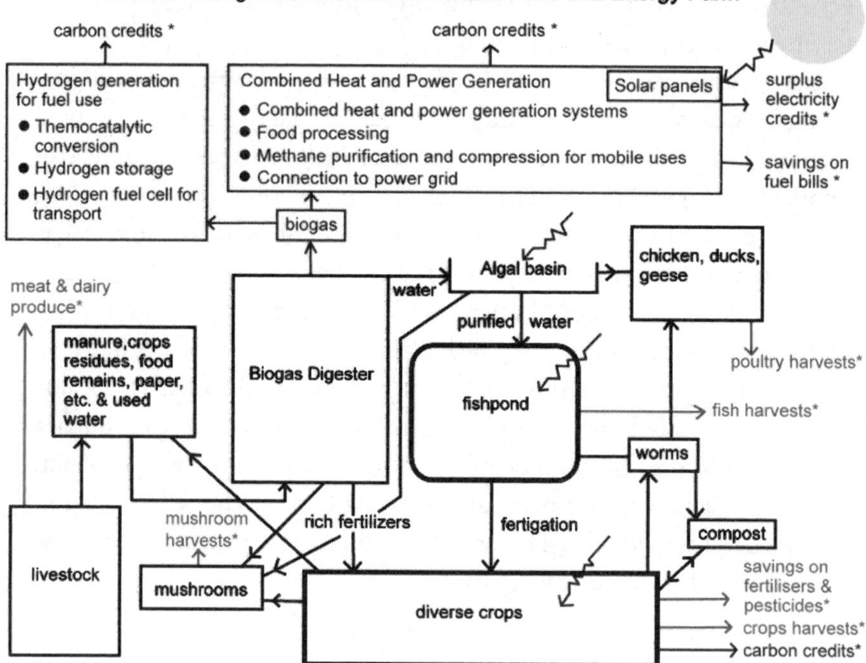

Table 1. Mae-Wan Ho's Dream Farm 2. A concept of an integrated 'zero-emission', 'zero-waste', highly productive farm that maximises the use of renewable energies and internal input, turning 'wastes' into food and energy resources, completely obviating the need for fossil fuels. Mae-Wan Ho says this model of an integrated farm has the properties of an organism. She developed Dream Farm 2 from George Chan's pioneering Dream Farm 1. (Inst.of Science in Society)

would heat up through entropy — waste energy. A closed system is one in which one organism's waste is put to use by another, so the energy generated is kept within the family, so to speak.

The problem we have today is of enormous waste of energy from farming systems, transport, building, manufacturing industry, power generation, which causes entropy and contributes to global warming, separately from the greenhouse gas excesses. Can you wonder why our environment is over-heating when the internal combustion engine wastes 70% of its energy, and the typical power station delivers only 40% of what is generates?

Biodynamic cultivation

Rudolf Steiner (1861–1925), a teacher and philosopher born in Austria, and founder of the anthroposophical movement, introduced biodynamic farming. Its approach to cultivation is similar to Schauberger's, which holds that energy is the primary cause, and growth the secondary effect. While it has been suggested that Rudolf Steiner and Viktor Schauberger did exchange ideas, it is not clear how much either might have influenced the other.

Biodynamic farming recognizes an ancient practice of burying cow's horns filled with cow dung deep underground in the autumn. At this season the spiral shape of the horn draws in the active Earth's energies, transforming its contents into powerful fructigenic energies by the cold process of fermentation encouraged by lower temperatures. The cow horns are disinterred in early spring, their contents having been converted into a sweet smelling, highly active substance.

This empowered material is the basis of the natural fertilizer known as '500 mix'. Since 1947, it has been increasing widely used, and over 1 1/4 million acres are fertilized in Australia using this system. The land where it has been spread, when seen from the air, stands out clearly from neighbouring farms, due to the much greener pasture. Some cows from farms bordering Alex de Podolinsky's did not eat for two or three days after they had broken into the biodynamic farm, so high was the quality of the grass they had consumed.[11]

The '500' mix fertilizer is derived from an ancient Alpine tradition which Schauberger himself once observed being practised by an old mountain farmer who achieved amazing results from his fertilizer. The principle is like that of homeopathy. When a homeopathic medicine is made, the original remedy is stirred and shaken between the dilutions, which increase its potency. To make the fertilizer a small quantity of the converted cow dung is added to water and stirred first in one direction and then in the other, so as to create alternating vortices rotating about the vertical axis of the mixing vessel.

A left-hand vortex builds up the positive energy and the right hand vortex creates a negative energy that draws in the inseminating O_2. The alternating energy charge builds up the inherent energies of the 500 mix. This recalls the alternating left and right hand bends in a river building up its energy in a longitudinal vortex (see p. 108).

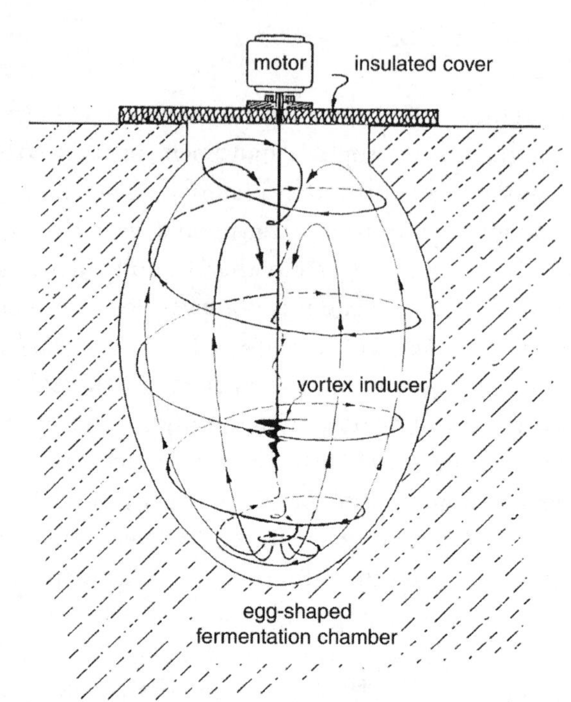

Figure 22. An egg-shaped fermentation chamber. This fermentation chamber is buried in the ground. The impeller induces a clockwise rotating vortex into the liquid, which allows the fermentation to draw in geospheric energy from the surrounding ground. (Callum Coats)

As the vortices are alternately formed and destroyed in the stirring, the level of energy rises and the degree of chaos decreases until, after about an hour, the product is ready for use. This is sprayed on the fields towards evening within two to three hours of preparation and before the accumulated energies have dispersed.

In order to learn to live sustainably, we have first to understand how Nature works, and accept her laws. Schauberger's life was devoted to this challenge.

The best guarantee that your food will be produced according to environmental and social principles is to meet the people that grow it.
Mae-Wan Ho, Food for the Future

19. Epilogue: The Big Picture

A free people can grow only from a free Earth. Any people that violate Mother-Earth have no right to a homeland, because high-quality races cannot survive in soils destroyed by speculation, that is, because they are divorced of all connection with the Earth. Human societies without roots perish. They have to experience the path of decay until, like unsuitable fertilizers, they give up their stubborn wills and only then will they be allowed to start again and re-enter the mighty course of evolution.

Viktor Schauberger, *Implosion* magazine, no. 37, p. 8

The holistic view

If your contact with the environment is limited to what you see from your window — your neighbour opposite and a view down the street — your understanding of the world will be minimal. With the instant global communication we have today compared to a century ago, you would expect our understanding of the world to be immensely greater. So how is it that we still find it so hard to see the big picture?

Psychologists tell us that the family is the best place to appreciate life's lessons, and to learn to get on with people of different temperaments, because there is a genetic requirement of mutual support and protection. The family is an organic unit that encourages holistic interdependence. Frequently dysfunctional family life in our 'each for himself', instant gratification society makes it hard to understand a holistic worldview. A genuine community is more holistic. Most religions in theory teach holistic behaviour, but to follow a way of life in which you value and appreciate every person with whom you come into contact may be idealistic to the point of sainthood.

If we see life systems only in mechanical terms, our worldview becomes blinkered and we are unaware of other levels of connection in the vast web of life.[1] It is the interconnectedness and interdependence of all of life that bring meaning to, and an understanding of, life and evolution. The prevailing worldview of Earth as a mechanism spills over into our need to rationalize every part of a process, a position which is uncomfortable with chicken/egg predicaments. The 'holistic' worldview is more accepting of mysteries, believing that they will gradually release their truth at the appropriate time.

We have talked a lot about holism in this book, but in reality it is an intellectual concept. Life is about experience, and it is quite difficult to *experience* the 'knowing' that everything is interconnected; the world's present values and our education are firmly prejudiced against any such reality. This is the stuff of mysticism. The transpersonal psychologist Abraham Maslow coined the term 'peak experience' for those extraordinary moments of insight when we see everything is actually one single reality.

My water epiphany came when I experienced for the first time that I am *embedded* in water. I saw water as the infinite continuum of life in which all living creatures are contained. This was much more powerful than the intellectual idea of being interconnected. 'We are all One', because of water.[2]

Other peak experiences may not seem to have a water connection, but I believe they often do, because it is water that enables personal experience and emotions, perhaps prompted by its etheric partner, the Divine plan. It might be a sense of wonder felt looking up at the infinite beauty of the night sky; a glorious sunset or a view from the mountaintop — or the singing of the nightingale; or perhaps an insight gained during deep meditation. Any one of these may trigger a very deep sense of longing in our soul, precipitating a change in our priorities and in our life's direction. We may find that we want to help other people or serve the planet.

The challenge is how to integrate such experiences into one's own life. They are a real gift, and to pass up the opportunity for change can be a source of real regret in later years. The 'highs' experienced in drug taking are not as real, because they are a 'dislocation' and incoherent with the whole organism (the 'soul'), seldom bringing such changes to the person involved.

I find the whole story of water quite remarkable, and very exciting, for clearly water is the common denominator of all life, implying that the Universe is a single organism, all its constituents being closely interdependent. One is reminded of Douglas Adams' phrase: 'the fundamental interconnectedness of all things'.

The search for meaning

For more than ninety-nine per cent of humanity's time on this planet we have been closely connected with Nature which, until modern times, was always considered sacred. The various gods that people worshipped were mostly thought to be part of the natural environment, while the shamanic view was often of the raw power of Gaia. Many religions (for instance, Celtic Christianity) were a blend of pagan belief and more modern monotheism.

With the Enlightenment the scientific revolution brought a rise of secularism and a polarization between religion and science. The Church, which had had a stabilizing influence on society, lost the support of many people who now considered its doctrines to be controlling and simplistic.

We have created God in our own image, a childlike conception which dilutes our own responsibility, and leads to division and to wars. A vacuum has been created in society, which has doubtless contributed to the breakdown in values and morality that we see today.

Humanity seems to be genetically wired to seek meaning in their lives at a personal level. Traditionally this has been exercised at the tribal or mass level. Mass worship is still the norm in religious groups. Increasingly, however, one sees the emergence of a more mystical and spiritual form of seeking, where the 'Big Daddy' version of the supreme intelligence/creator is transcended, yet allowing a personal connection with a higher level of consciousness, with more emphasis on contemplation and meditation rather than worship.

One of the by-products of the Enlightenment has been the utopian belief that increasing use of technologies could bring about transformation of society. This encouraged the 'quick fix' attitude rather than the longer view, which has dominated our policies towards problems, from financial policies to global warming, to Afghanistan. This is the mark of a reductionist rather than an holistic world view.

Holistic science, especially at the quantum level, can demonstrate a level of meaning in all life's processes, and I can foresee a new spiritual consciousness arising as a result of people rediscovering the sacred in Nature and welcoming the idea of a universal intelligence behind creation, uninfluenced by male anthropomorphic constructions. Indeed, an appreciation of water's part in life is, I believe, the key to a new spiritual worldview.

Holism in society

We are mistaken to separate human society from the natural environment. As part of the planetary community we are subject also to Nature's laws. As the crunch point has arrived for Nature, so it has for our human institutions. If they are not sustainable, they will fail. There is a strong link between sustainability and holism, for both are connected to water's qualities.

Water is the common denominator of all of life, so to understand life we need to understand water and the sustainability it requires of life. Water is the medium of communication within and between organisms. It is through water that the complex network of life is organized, creating interdependence between all life forms.

Water is also the model of holism. Through studying water's qualities we can learn how to create sustainable enterprises and communities, where energy produced by one part can be recycled by another (little circles within the whole), instead of being wasted and causing entropy.

It was Mae-wan Ho who proposed the model of energy sustainability for organisms which operate efficiently with minimum waste and entropy.[3] Entropy is merely dissipated energy that is useless for doing work but which instead heats up the environment.

As we saw in Chapter 18, she applies the same model of the zero waste and zero entropy to the Dream Farm, where all parts of the system feed each other, keeping the energy 'in the family'.[4]

An example for the sustainable community that fits Ben Okri's wish-list is that of the Transition Movement, which was formed in 2006 by

Rob Hopkins in response to the twin challenges of diminishing oil supplies (Peak Oil) and climate change.[5]

Most of us avoid thinking about what will happen when the oil gets short or becomes too costly, but the Transition movement shows how the profound and inevitable changes ahead can have a positive outcome if we are prepared. A Transition town is one that has taken steps to harness the energy and enthusiasm of people of all ages to recreate a cohesive self-reliant community with a resilience that it lost with the coming of the oil glut sixty years ago; a community that will grow its own food, generate its own energy, use local skills and creative talents, care for its people and give everyone a sense of engagement and purpose. I strongly recommend that all who read this should join this creative movement. You will find in it a recipe for increasing optimism in a future for humanity.

There are nearly a hundred Transition Towns in the British Isles, and some five hundred more communities aspiring to become one (they have to learn how), and the movement is spreading in Australia, New Zealand, North America and Europe. It is a holistic movement in which controlling personalities are discouraged, and everyone is encouraged to play a part.[6]

'Sustainability' is a word that is fashionable, but often misunderstood. If taking the long view is unfamiliar to you, you probably don't understand the concept. It is more often a rationalization than a heart-felt instinct. The best definition is Mae-Wan Ho's: to think of sustainable systems as organisms, in which all energy is kept within the whole, resulting in zero waste and zero entropy. Surplus energy from one part of the organism is used by another part.

Our present economic systems are basically unsustainable. Interestingly, the Transition towns of Totnes and Lewes in southern England, by persuading the town's stores to work with a local currency are reducing the amount of money energy that is taken out by national businesses. There are local currencies, too, on the West Coast of the US.

The most important lesson is that Nature will not tolerate gross inequalities: water's role is to bring down to size, to level, to stimulate wellbeing of the whole, not individuals in isolation. Many now believe that it is the level of inequality that contributes more to our social ills and personal disaffections than any other factor.[7]

The quantum view vindicates traditional values and what are sometimes called 'natural laws'. Mae-Wan Ho made the extraordinary suggestion that the way coherent organisms become entangled with one another is the prerequisite for universal love and ethics.

Water's fundamental property is that of holism (the interdependence of all life), which was the model for human society until the human ego took over. This model is that of the ideal integrated community and of true sustainability.[8]

Sacred chaos

The turmoil of 2008–2010 in the workings of the capitalist system — banking, credit, the markets, both local and international, globalization of commerce — have shown that our financial systems are unsustainable. The prevailing economic model of unlimited growth confuses money (whose role is exchange) with real wealth (production of goods and services).[9] It produces waste and entropy, compared to a natural system where the energy is recycled within the system. The concept of true wealth being about the wellbeing and happiness of society and the wellbeing of the whole planet is not widely accepted today. We are all implicated, having been seduced by the mirage of material wealth.

Until there is a change of our worldview of the purpose of human society, don't expect our social and economic problems to go away.

The creative partnership

It is clear that organisms, even across the genera boundaries, are aware of each other. This awareness very likely comes through a shared grounding in the water element. On the micro level these are mostly electro-magnetic relationships which are governed by polarities. These polarities are the dynamos of life, creating the great dialectic principles that also carry complementary attributes, like positive/negative, light/darkness, good/evil, heat/cold, a balance between which is necessary for an understanding of the whole.[10]

Because the etheric field is universal and water is common in the Universe, associated with life or potential life, it seems natural to view them as complementary, working together. In esoteric terms the etheric field (sometimes termed 'the God field'), may be seen as the masculine, initiating polarity, while water has the receptive, feminine role.

The idea is found in many religions that God cannot act directly on the world stage, but only through incarnate life, which may be the reason for periodic visitations by great teachers to help correct imbalances that inevitably arise where free will is involved.

What water can teach us

—On a physical level, we are composed of at least 70% water. Therefore, just for health reasons, we should be aware how water quality affects us physically.

—Water is associated with the emotions and with creativity. Becoming more aware of water's dynamic energy will enrich our emotional and creative life.

—Water is the common denominator of all of life, so to understand life, we need to understand water and the sustainability it requires of life. All of life is linked by water — it is what gives humans a sense of connection with Nature and with other animals.

—Water moulds the landscape, controls the weather and climate, growth and fertility, health and sickness, harmony and beauty, communication between organisms, pulsation and rhythm. Understanding how this works will help us to be better guardians of the Earth.

—The most important lesson is that Nature will not tolerate gross inequalities: water's role is to bring down to size, to level, to stimulate fertility and productivity of the whole, not individuals in isolation.

—Humanity's growth has reached the point of smothering the rest of creation and, with its gross inequalities between rich and poor, in individual countries and between nations, human society has failed the test of a sustainable organism.

—Through water, quantum coherent organisms invariably become entangled with one another. A quantum world is a world of universal mutual entanglement, the prerequisite for universal love and ethics. Because we are all entangled, and each being is implicit in every other, the best way to benefit oneself is to benefit the other.
—Thinking like water, our personalities are able to soften, so that we can empathize better with others. Understanding why water flows the way it does will teach us that it is better to 'go with the flow'.
—An important role of water is to carry an electric charge, which it can do only when impure and unstable. Water teaches us that impurity and instability can stimulate creativity and growth.
—Water adapts itself to the shape of its container. If we are able to adapt to a situation that is presented to us, we will be better able to change negative situations from within.
—Its balance of flexibility and stability is a metaphor of physical reality and ideally of our own relationships with ourselves and others.
—In the way it generates equilibrium and harmony, water is a brilliant ecological model. Its movement and rhythm cultivate the beauty of an aesthetic model.
—Water's fundamental property is that of holism (the interdependence of all life), which was the model for human society until individualism became rampant, the ego inflated, and when the lust for (seduction by) wealth and power became dominant. This earlier model is that of the ideal integrated community and of true sustainability.
—Solubility is the capacity to assimilate anything and everything into a context where all can coexist. This essential property of water provides us with a model of the ideal integrated community.
—Humanity's present identification with a mechanistic worldview and the falsity of our present human social and economic structures make a holistic worldview impossible; water can show up our denial of the obvious.

—It is the restlessness of water which gives living systems the ability to become ever more complex and to strive towards the perfection they can never reach. As the driver of evolution, water is a model for our own striving, allowing us to evolve through our own mistakes. This is the self-empowerment process of holistic biology.
—Water raises its dynamic energy through turbulence and chaos which, in the bigger picture, mirrors the turbulence of change in human society.
—The dynamic way that water creates energy shows up our technology's way of generating energy as wasteful, dangerous, and harmful to the environment.
—Perhaps an appreciation of what water contributes to the collective consciousness may help us better understand our own heritage, our past experiences and our life purpose.

Water and consciousness

The quantum field, the enormous web of interconnected subtle energy that seems to continue infinitely through space is a kind of communication system, making all of creation aware of itself and its one-ness. Viktor Schauberger called it 'the Eternally Creative Intelligence', 'The-All-That-Is'; it is the foundation of consciousness, also known as the 'God Field'.

Some scientists call it the Zero Point Energy Field ('the Field' for short), because changes in the field are still detectable at absolute zero when all matter has been removed and no motion is possible. It is now recognized that our concepts of space and time are relevant only at the material and not at the dynamic energy level.

We are clearly at the end of one epoch, and embarking, albeit hesitatingly, on a new, uncertain one. The new quantum physics has started to rekindle ideas of a universal consciousness and spirituality. Science is, if you like, proving the existence of a consciousness-like 'God' or Eternally-Creative Intelligence; and that the Universe does actually have meaning and probably even a purpose. But there is a lot about the traditional knowledge of the hierarchy of energies and dimensions, familiar to esotericists, that the new generation of quantum physicists still has to take on board.

An annual conference is held in London called 'Living the Field: the Marriage of Science and the Spirit', bringing together scientists like David Bohm, Deepak Chopra, Brian Goodwin, Karl Pribram, Rupert Sheldrake and Russell Targ, to inform us of these startling discoveries.[11]

To me, the 'Zero Point Energy Field' has a certain barrenness about it; a masculine mental concept. It seems a bit cold and 'out there'. I feel that it can be brought to life and closer to home with the concept of water as the medium of universal consciousness. After all, we live on such a beautiful planet full of colour and diversity, made vibrant because of living water.

Cosmic correspondences

Viktor Schauberger demonstrated that Nature's dynamic is the Law of Polarity. He said that unity or the whole is composed of two opposing qualities in balance. The dance of creation is the harmonious interplay through attraction and repulsion of polarized atoms, without which there would be no water, no plants, nor chemical compounds. The mutual attraction of 2x H and 1x O gives birth to the marvel of water. He showed how the Sun, as inseminator of life, carries a positive (masculine) charge, while the Earth's energy is receptive and feminine.

Turning around the principle 'as above, so below', it is reasonable to expect that the ultimate Cosmic dynamic of life might be such a polarity. Substitute for the Sun the Etheric Field and for the Earth the Domain of Water, and a whole list of correspondences appears that demonstrates how, even at the cosmic scale, reality and wholeness come with the resolution of the polarity of qualities. The etheric field is the initiator, the water domain the implementer.[12] Neither can work without the other. That is why I call water the handmaiden of the etheric.

Complementary Attributes

The Quantum/Etheric Field	The Water Domain
Sun	Moon
the god field	Earth/ Nature
the Divine plan	evolution
positive/male/*yang* energy	negative/feminine/*yin* energy
will	consciousness
order	chaos
laws	procedures
initiating	following thru
impulse	implement/fulfil
creating	sustaining/nurturing
formative	relatedness
detached	intimate
intellectual	emotional
theory	practice
idealism	beauty
determination	appreciation
goal orientation	acceptance
planning	going with the flow
implicate	explicit

Water as spiritual mediator

This takes us back to what I said at the very beginning, about our collective amnesia and not living up to our potential.

From their inherited traditions, we know that our ancestors, who were self-aware and imaginative, believed in the divine nature of water, and had a sense of how it connected all of life — that it is, in fact, a medium of universal consciousness. Even the surviving fragments of dysfunctional indigenous societies of more recent times acknowledge this.

Schauberger and Steiner both said that Earth's water acts as the mediator or a transducer between the solar and cosmic energies on the one hand, and the energy of the Earth and Nature on the other, to produce productive, balanced energy for the emergence and evolution of conscious life, for all life-forms and life processes.

My personal vision is that living water is the key to evolution, to healing and to consciousness. Jesus was baptized with living water. We should venerate it, as indeed did our forebears. Having examined the extraordinary roles of water in the initiation, maintenance and evolution of life, it is not hubris to say that water is the creator of life, for I believe it is the active part of the Universal Creative Intelligence.

We need to have more faith in our instincts. Many of us may have a dim memory of when we were part of Nature. This is in our blood — or should I say, in our water memory! We must tell our children that the Universe is a single unity, that all of Nature is one. We have been living under an illusion of separation.

In times like these, when many people are worried about their security, their jobs and their savings, it is helpful to have a broader view of what life is really about. To contemplate the astronauts' view of the shining blue pearl of Earth, or to gaze up at the resplendent night sky, has a way of putting things in their true perspective. Our individual concerns diminish in the face of the unity of life. This is what true holism is about.

We are now beginning to witness a paradigm change in our attitude to all of life and to our worldview, to which an understanding of water as a medium of consciousness contributes substantially. Consciousness change happens at a personal level. Eventually the change in worldview will trickle down to mainstream science, but let's start with you and me.

Appendices

1. Water and Health

Water plays an essential part in health and healing. It is the source of either good health or of sickness. Our bodies literally run on water, so we ought to be drinking the best quality water for our health. In developed countries public water supplies are often unhealthy. They come mostly from underground sources, often well laced with agricultural chemicals, heavy metals and other toxins.

Many of the physical pollutants are filtered out by water authorities, but at the price of adding chlorine, fluorides or other chemicals which can damage our bodies in the long term. In addition, the subtle energy of these pollutants are not removed by physical filtering and have to be dealt with by a quantum technique (for instance, Nature's method — the vortex).

Tragically, 35 per cent of the world's population does not have access to drinkable water, never mind good quality water.

The qualities of different waters

Although good water is tasteless, without colour or smell, it quenches our thirst like nothing else. In order to be healthy, we need to drink, according to most authorities, 5–9 pints (1–2 litres) of good quality water a day. Some types of water are more suitable for drinking than others. High quality water should contain elements of both earth origin (female) and atmospheric (male).

Distilled water

This is considered physically and chemically to be the purest form of water. Its nature is to extract or attract to itself all the substances it needs to become mature and, therefore, absorb everything within reach. Such water should not be drunk every day. The 'Kneipp cure' uses distilled water for its short-term therapeutic effect, where it acts to purge the body of excessive deposits of particular substances. There is controversy about its safety. Some medical authorities justify the drinking of distilled water by asserting that only the unusable minerals are leached out of the body. I should advise caution!

Rainwater

As long as it has not been affected by industrial pollution (acid rain), rainwater is the purest naturally available water. Though slightly richer through the absorption of atmospheric gases, like distilled water it is still unsuitable for drinking over the long term. When drunk as melted snow-water, it can also give rise to certain deficiencies; and if no other water is available it can on occasion result in goitre, the enlargement of the thyroid gland.

Juvenile water

Juvenile water is immature water from deep underground sources, like geysers. It has not mellowed sufficiently on its passage through the ground. It has not developed a mature structure and contains some minerals (geospheric elements), but few gases (atmospheric elements), so as drinking water it is not very high grade (cf. most spa waters which arise from mineral rich depths).

Surface water

Water from dams and reservoirs contain some minerals and salts absorbed through contact with the soil and the atmosphere. Its quality deteriorates through exposure to the Sun, to excessive warming and to chemicals and other pollutants. Although most urban communities now depend on this source, generally speaking it is not good quality water.

Groundwater

Groundwater has a higher quality due to a larger amount of dissolved carbones[1] and other trace salts. This is water emanating from lower levels, seeping out at the surface after passage along an impervious rock surface. Often this is now polluted by the chemicals of industrial agriculture.

Spring water

True spring water has a large amount of dissolved carbones and minerals. Its high quality is often shown by its shimmering, vibrant bluish colour. The product of infiltrating rainwater (with a full complement of atmospheric gases) and geospheric water (full complement of minerals, salts and trace elements), this is the best water for drinking, and it often retains this quality in the upper reaches of a mountain stream. Commercially bottled 'spring water' is unfortunately not always of the best quality, even if it is bottled in glass rather than the plastic that impairs its quality; and many are not from true springs.

Other groundwater

Aquifer water is obtained from boreholes and is of unpredictable quality. It may be saline, brackish, or fresh. Water from wells can vary from good to poor, depending on how deep is the well and what stratum of water is tapped, and they can be polluted by nitrates and herbicides, but they can be purified with a vortex filtration system.

Water supply

It was not until the nineteenth century in Europe that priority was given to building sewers to remove waste. Before then, public hygiene was not a high priority. There was little understanding of the connection between water and health.

The Romans knew about keeping water healthy: they made wooden water pipes to let the water 'breathe'. However, they also drank wine out of lead tankards! Pipes that were made of lead caused much illness for centuries. More recently, iron or steel pipes were lined with aluminium to prevent rust. It is no wonder that we still have a metal toxicity problem with our water supply that requires specialized filtering. Make sure that the filter you use takes out heavy metals (see below).

Bottled water

The United Kingdom spends £1.85 billion per year on bottled water, but the 43 litres per head (2007) is small compared to the west European average of 125 litres each or the US rate of 100 litres. With an average price of £1.25 a litre, it is more expensive than gasoline (petrol), even with its heavy tax. People seem to believe that bottled water is healthier than what comes from the tap, but this is not necessarily true. There are much stricter regulations with water purity for tap water than there are for bottled.[2] There is little evidence that bottled water is healthier, but this must depend on the tap water being compared. However, it is easy to filter and re-energize tap water to make it healthy and drinkable, and this need not be costly.

The health scares with some of the better known brands, such as benzene found in Perrier water in 1989, bromate in Dasani in 2004, and naphthalene in Volvic in 2005, and the occasional appearance of nitrates don't seem to have put people off. The temptation to increase the shelf life and make the water taste more interesting, has led to the addition of preservatives and flavourings (not listed on the label), which usually come from petrochemicals, and often include neurotoxins, carcinogens, benzoates and artificial sweeteners. Many of the brands have a high mineral content which, over time, can put a strain on the kidneys (see *The Ecologist,* September 2007). Patrick Flanagan even claims that much commercial bottled water in the USA is actually tap water.

Perhaps more obviously, but seldom considered, is the fact that chemicals can leach out of the plastic into the water, particularly in strong light or warmer temperatures, or especially with re-use. Plastic bottles do not easily break down or recycle, so they are a major environmental problem, whether in landfill sites, on beaches or in the ocean (for instance, the North Pacific gyre) and produce significant pollution when burned.[3]

We are not saying that people should not drink bottled water, but be careful, choose a glass bottle, and don't assume it's always good quality. It is very convenient to drink it from time to time, but to depend on bottled water could also make you anxious when it runs out. We prefer to carry our water in a metal screw-top jug when we go out.

What we do recommend is to purchase a good water filter, preferably plumbed-in, to remove the physical pollutants (you should be aware of what it does and does not remove) and something to re-energize and restructure the water to deal with the energies of those pollutants that the filter can't remove. The cost is a fraction of bottled water.

Water Purification

We are all concerned about the quality of our drinking water. We hear that in the bigger cities, municipal water is recycled, in London and New York as many as twenty times. But it is filtered, isn't it? So it must be safe. Yes, it is filtered, but that doesn't remove all the germs, so chlorine is added. Viktor Schauberger did a lot of research on the effects of chlorine. We looked at chlorination and fluoridization in Chapter 7.

Alkalinity

A water ionizer separates alkaline from acidic water by electrolysis. The minerals with negative ions (for instance, magnesium and calcium) are attracted by the positive electric anode, making that part of the water alkaline, while the negative cathode attracts the acid part of the water.

Alkaline water has a high oxygen reduction potential (ORP) which can neutralize free oxygen radicals, the source of much disease and premature ageing. Many chronic diseases are encouraged by excess acidity in the body which often stems from poor diet and lifestyle.

Conventional medical opinion is sceptical of the benefits of drinking alkaline water. It would be unwise to make generalizations about its possible benefit, because the reaction must be individual. Too much alkaline water is undoubtedly harmful to some; it must be a question of moderation and balance. Taking it between meals when the stomach acids are less active would, for example, be preferable. (Alkaline water made from bicarbonate does not have the all-important ORP.)

Distillation

Distillation is an ancient process of vaporizing water so that the pure water molecules are separated as steam from the contaminants which have a higher boiling point. The steam is allowed to condense through tubes into another container. The distillation process removes minerals, viruses and bacteria, and any chemicals that have a higher boiling point than water.

It has been used for centuries in the making of whisky and other spirits, and in the 1970s was a popular method of home water purification. Distillation is often used as the preferred water purification method in developing nations, or areas where the risk of waterborne disease is high, due to its unique capability to remove bacteria and viruses from drinking water.

However, there are reasons why distillation should not be used for creating drinking water. It does not remove chemicals which have a lower boiling point than water, such as chlorine or its by-products, and volatile organic chemicals (VOCs) like herbicides and insecticides which have a lower boiling point than water. Municipal treatment of water removes bacteria and some heavy metals, but does not remove VOCs; nor does distillation, which many who still use it mistakenly believe.

The mineral-free water produced by distillation is acidic, and can be quite dangerous to the body. Acidic drinking water can dissolve the essential mineral constituents from bones and teeth. In addition, distillation is incredibly wasteful. Eighty per cent of the water is removed, discarded with the contaminants, which may be alright with sea water, but less justified with precious fresh water, leaving only one gallon of purified water for every five gallons treated.

Distilled water has an important use in scientific experiments and in some industries which require mineral free water. It removes heavy metal materials like lead, arsenic and mercury from water and hardening agents like calcium and phosphorous.

Viktor Schauberger was adamant that drinking distilled water was undesirable. He called it 'immature' water which is aggressive and 'hungry' and can be destructive. When its energy is raised through vorticizing, this unpredictable quality can be magnified (like a young child's). This is why we advise against trying to restructure it through vorticizing. (Distilled water can't be ionized.)

Reverse osmosis

Reverse osmosis is a modern system for purifying water. It was developed about 40 years ago as a treatment for desalinating seawater.

It seemed to be the answer for water purification at home, and was popular in the 1970s as a cheaper alternative to distillation.

In natural osmosis water tends to migrate through a semi-permeable membrane from a weaker to a stronger saline solution, balancing the saline composition of each solution. The reverse osmosis process also employs a semi-permeable membrane, but water is forced through it under pressure.

The membrane blocks the passage of salt particles which are physically larger than water molecules. It will also remove larger particles of other contaminants: lead, manganese, iron, calcium and the dangerous additive fluoride that is sometimes added by municipal authorities, so it is widely thought to be the answer to water purification.

However, there is a downside to this. Reverse osmosis does not remove the smaller sized particle contaminants such as the VOCs, for instance, chlorine. Like distillation, reverse osmosis, by removing alkaline mineral constituents of water, produces acidic water. It can be quite dangerous to the body and may dissolve calcium and other essential mineral constituents from bones and teeth. The trace elements of minerals are essential constituents of fresh water; their removal leaves drinking water tasteless and unhealthy. Taking medicinal supplements of the minerals you need to replace is not a convincing solution, for they are not delivered in the balanced form the body requires.

Reverse osmosis, although it is less wasteful than distillation, is still a most inefficient process. On average, it wastes three gallons of water for every one gallon of purified water it produces. Reverse osmosis was not used in Schauberger's day, but his criticisms of distillation would apply to osmotic filters.

Water filters

There is a wide variety of drinking water filters available; the expensive ones are not necessarily the best. It is important to consider a drinking water filter purchase very carefully. There are a number costing $340 (£200) and more, which are no better than others selling for under $170 (£100). You need to know the quality and content of your water to choose the appropriate filter with detailed information about the filter's applicability. (See Links and Resources.)

You should look for a filter that removes bacteria and suspended solids, heavy metals, chlorine and chemicals and dissolved organic matter. The better brands offer alternative cartridges for specialized problems like fluoride, excessive agrochemicals and calcium.

The basic filter jug

This uses granulated carbon. It removes chlorine, some chemicals, mercury, large parasites and particles. *Advantage:* cheap initial cost. *Disadvantages:* short cartridge life; won't remove bacteria, some heavy metals, asbestos, radioactive material.

Plumbed-in filters

These come with separate sink tap and cartridges (most are easy to install yourself):

Carbon block and activated carbon filters remove what the basic jug does, and also dissolved organic matter such as pesticides and other chemicals, oil residues and some radioactive substances. *Advantages:* usually relatively good value, long lasting cartridge. *Disadvantages:* some of these do not remove heavy metals, fluoride or viruses. However, some water filters offer a combination of cartridges which may include: removal of fluoride, nitrates, excess calcium or iron and heavy metals.

Ion exchange cartridge removes most heavy metals. It is often incorporated in a general purpose cartridge.

Ceramic (often silver coated for antibiotic results) removes finer particles than the carbon does. A ceramic filter is normally more expensive.

Negative ionization makes water more alkaline. It is widely believed that acidity produced by processed foods, drugs, stress and pollution is a major cause of illness. We take the view, however, that eating more alkaline food (and less acidic) is a more successful way of balancing the body's pH than through expensive water ionization.

Magnetizing water: The main use of magnets with water is to break up the normally tight structure of the water molecule. This allows the body to absorb a greater amount of oxygen and hydrogen. Magnets will also reduce the tendency of hard water to deposit chalky sediment. Magnets are also used to increase the efficiency of fuel oils.

Distillation and reverse osmosis: for the reasons given above, we do not recommend these methods for drinking water, unless there is a specific medical need for immature, pure water. We favour retaining the minerals, salts and trace elements as far as possible to maintain the water's 'maturity', while removing the dangerous metals and chemicals. It is really not practical to replace the minerals artificially in distilled or osmotic water.

Shower filter: Most people shower these days. Unless you're a 'shower-hogger' they take much less water than a bath. In a shower we absorb the chlorine other chemicals in the water through our skin. It also evaporates as chloroform, which can make us drowsy. A shower filter will greatly reduce these effects. They need not be expensive or reduce the pressure, and should improve alkalinity. If you don't want to plumb-in a whole house filter, this might be a good idea.

Whole house filter: Point-of-entry plumbed-in filters will provide filtered water to all baths, showers and washing machines. When you filter all of the water entering your home, you improve not only the healthfulness of the water, but the indoor air quality as well, as it stops chemicals vaporizing (for instance, from your clothes in the washing machine). It reduces the risk of respiratory problems.[4] (See Links and Resources.)

A whole house filter need not be a big investment. Do shop around, because the expensive ones may not be any better than the reasonable ones.

Imploded water energy harmonizers: Produced by the Centre for Implosion Research (CIR), these are by-products of Viktor Schauberger's technology, using highly energized water spiralling in copper tubes formed into beautiful organic sculptures. The CIR has energy harmonizers for both personal and environmental protection from electromagnetic pollution. One type will clear algae growth or reduce calcification of water pipes.[5]

Water restructuring

Nature does not destroy — it recycles. Nature purifies water by means of the vortex (see Chapter 12), which raises the dynamic energy level of the water to that approaching, or higher than, our own intercellular water. This is the best form of water to drink. It means that the body will have to work less hard to produce quantum water, and will maintain a higher level of subtle energy in the body.

Some products claim that 'Water spun anti-clockwise (left spin) is harmful' (that is, 'negative'). This is incorrect. Nature's way of raising energy is to induce an oscillating movement alternating between negative *(yin)* and positive *(yang)* electromagnetic qualities. (See Introduction.) Each change of direction of spin raises the energy slightly, with cumulative effect. With energies, the term 'negative' refers to polarity, not to quality.[6]

There are some filters that include a vortexing function, but they are expensive and may not do what you want. What we do for cooking or drinking is to pour the water from our plumbed-in filter into either a vortex jug or a double-egg vortexer.

Health aspects of living water

The essential requirements for living water are that it should be:

—*clean:* free of biological or chemical contamination and harmful energies (which increase oxidization and free radicals)
—*mineralized:* especially magnesium, calcium and trace minerals, which will produce greater alkalinity
—*with smaller or looser molecular clusters:* (micro water or magnetized water) gives low surface tension — and easier absorption
—*with abundant negative hydrogen:* produces a higher anti-oxidant potential
—*energized:* with positive, beneficial etheric energies, best done through vorticizing.

Water purity

Many filters remove biological and chemical pollutants. The best are the plumbed-in types. Avoid distillation and reverse osmosis, because these act like antibiotics, removing the good as well as the bad. It is hard to replace the minerals lost, and their use can de-mineralize the body.

Polluting energy

The energies of hospital wastes, hormones and estrogen are not removed by filters, but can be neutralized through vorticizing.

Humidifiers (and negative ionizers)

Many people overheat their homes with central heating, which can make the indoor air too dry and unhealthy. An electric humidifier can be combined with the use of bactericidal essential oils to stop the spread of infection from people with colds or bronchitis. (A negative ionizer will also improve the indoor environment, especially in hot weather.)

Minerals

Water normally supplies a substantial part of the minerals we need. Many of our soils now suffer from mineral depletion, so that we receive less nutrition from our food. We would be wise to take as supplements those minerals of which we are deficient (hair analysis is usually reliable).

Hydrotherapy

The tradition of swimming and bathing goes back millennia, sometimes as a ritual for spiritual purification, but also very much for health. The spa culture was a Roman thing, and often the focus of fashionable social life. Warmer water is particularly beneficial for sore bodies, arthritics and physical recuperation.

Thalassotherapy is a specialized variation practised with sea water, popular on the west coast of France, and on the Dead and Red Seas. It is considered beneficial for skin and circulatory disorders, as well as joint stiffness and arthritis.

Near Eastern countries developed their own variation of the Roman baths. The Turkish bath involves a steam room and a sweat room, followed by a cold plunge and a massage. The Scandinavian sauna is a steam bath often followed by a cold douche.

Inhalation

This is an effective way of helping sinus or head cold symptoms. You put a few drops of an appropriate essential oil on the surface of a bowl of just boiled water, and put a towel over your head. Alternatively an electric face steamer is a worthwhile purchase.

Colonic irrigation

This is a detoxifying therapy popular in America and Britain, a cleansing of the colon with a pressurized water enema.

Using Water at Home

You may live in a part of the country used to having plenty of rainfall, so why worry about water use?

Rainfall patterns are becoming less predictable, our population is growing and our lifestyles are changing — we use 70% more water than we did forty years ago. In Southern California investment is now being committed to distilling fresh water from the ocean. In South-east England (Kent), water supplies are so critically short that using nuclear power to distil sea water is being considered.

Water meters are becoming more common, with the incentive to reduce needless use. It is the same the world over. Fresh water will become increasingly scarce.

How much water do you use?

The big users are:

A bath = 80 litres
Five minute power shower = 90 litres
Five minute ordinary shower = 35 litres

Washing machine = 60 litres
Dishwasher = 40 litres
Hosepipe or sprinkler = 540 litres/hour

Some tips on water economy

—Never leave a tap running. A dripping tap can waste more than 5,500 litres a year.

—Store cold water in the ice-box (fridge) rather than waiting for the tap to run cold. Cold water has greater dynamic energy.

—When you make a cup of tea, fill the kettle only as much as you need.

—By installing dual-flush or slimline toilets, buying water-efficient appliances and using low-flow taps you can easily reduce your water consumption by 25%.

—Consider installing a whole house dechlorinating filter, which will make your home environment much healthier.

—Save even more by doing your dishes in a bowl in the sink; the same with a few clothes. Half-load washing machine programmes are uneconomic on both water and energy.

—In temperate climates, the average roof collects about 85,000 litres of rain a year, enough, to fill 450 water butts with free water. Use this for watering your garden rather than wasting treated drinking water. It's a good plan to have a second butt to take overflow from the first.

—Use a watering can for your plants. A hosepipe or sprinkler uses a lot of water, and is often illegal anyway.

—Water your plants early in the morning or late in the evening to reduce evaporation. Direct water to the roots of plants and give then a good soak, twice a week is enough, even in warm weather.

—You may use grey water (from sinks, dishwasher, bath, washing machine) on your flowers, but preferably not on your vegetables.

—Group your vegetable according to their water need: leafy ones require the most, marrows and cucumbers when their fruit begins to swell; root vegetables require the least.

—Build up the organic content of your soil by using lots of compost: instead of throwing away your garden waste, potato peelings, paper and cardboard, compost them instead.

—Mulch on the surface of the soil will keep the soil moist. Strips of cardboard or carpet will also keep down the weeds. The less you dig, hoe and disturb the soil, the more efficiently will the organically rich humus thrive and retain the fertility and moisture of the soil. Permaculture is the most water-efficient method of gardening.

2. Water Anomalies

(Summarized, with permission, from Martin Chaplin: www.lsbu.ac.uk/water/anmlies.html)

Water is an apparently simple (H_2O) with a highly complex character (see Plate 3). As a gas it is one of lightest known; as a liquid it is much denser than expected (in the chemical group of hydrogen compounds); and as a solid it is much lighter than expected. Much of the behaviour of liquid water is quite different from what is found with other liquids, giving rise to the term 'the anomalous properties of water'. It is quite extraordinary that these anomalies have all enabled life to prosper on the Earth.

As liquid water is so commonplace in our everyday lives, it is often regarded as a 'typical' liquid. In reality water is most atypical, behaving as a quite different material at low temperatures to that when it is hot. It has often been stated that *life depends on these anomalous properties of water.* In particular, the large heat capacity, high thermal conductivity and high water content in organisms contribute to thermal regulation and prevent local temperature fluctuations, thus allowing us more easily to control our body temperature. The *high latent heat* of evaporation gives *resistance to dehydration and considerable evaporative cooling.*

Water is an excellent solvent due to its polarity, high dielectric constant and small molecular size, particularly for polar and ionic compounds and salts. It has unique hydration properties for biological macromolecules (particularly proteins and nucleic acids) that determine their three-dimensional structures, and hence their functions, in solution. This hydration forms gels that can reversibly undergo the gel-sol phase transitions that underlie many cellular mechanisms. Water ionizes and allows easy proton exchange between molecules, so contributing to the richness of the ionic interactions in biology.

At 4°C water expands on heating *or* cooling. This density maximum together with the low ice density results in (i) the necessity that most of a body of fresh water (not just its surface) is close to 4°C before any freezing can occur; (ii) the freezing of rivers, lakes and oceans is from the top down, so permitting survival of the bottom ecology, insulating the water from further freezing, reflecting back sunlight into space and allowing rapid thawing; and (iii) density-driven thermal convection causing seasonal mixing in deeper temperate waters carries life-providing oxygen into the depths.

The *large heat capacity* of the oceans and seas allows them to act as heat reservoirs so that ocean temperatures vary only a third as much as land temperatures and so moderate our climate (for example, the Gulf Stream carries tropical warmth to northwestern Europe). The *compressibility of water* reduces the sea level by about 40 metres giving us 5% more land. Water's high surface tension plus its expansion on freezing encourages the *erosion of rocks to give soil* for forests and our agriculture.

Notable amongst the anomalies of water are the *opposite properties of hot and cold water,* with the anomalous behaviour more accentuated at low temperatures where the properties of supercooled water often diverge from those of hexagonal ice. As a cold liquid water shrinks when heated. It becomes less easy to compress, its refractive index increases, the speed of sound within it increases, gases become less soluble and it is easier to heat and conducts heat better.

In contrast, as hot liquid water is heated it expands, it becomes easier to compress, its refractive index reduces, the speed of sound within it decreases, gases become more soluble and it is harder to heat and becomes a poorer conductor of heat. With increasing pressure, cold water molecules move faster but hot water molecules move slower. Hot water freezes faster than cold water, and ice melts when compressed except at high pressures when liquid water freezes when compressed. *No other material is commonly found as solid, liquid and gas within such a narrow temperature range as life on Earth requires.*

3. The Moral Bankruptcy of our Civilization

Ben Okri

The crisis affecting our economy is a crisis of our civilization. The values that we hold dear are the very same that got us to this point. The meltdown in the economy is a harsh metaphor of the meltdown of some of our value systems. For decades poets and artists have been crying in the wilderness about the wasteland, the debacle, the apocalypse. But apparent economic triumph has deafened us to these warnings. Now it is necessary to look at this crisis as a symptom of things gone wrong in our culture.

Individualism has been raised almost to a religion, appearance made more important than substance. Success justifies greed, and greed justifies indifference to fellow human beings. We thought that our actions affected only our own sphere, but the way that appalling decisions made in America have set off a domino effect makes it necessary to bring new ideas to the forefront of our civilization. The most important is that we are more connected than we suspected. A visible and invisible mesh links economies and cultures around the globe to the great military and economic centres.

The only hope lies in a fundamental re-examination of the values that we have lived by in the past 30 years. It wouldn't do just to improve the banking system — we need to redesign the whole edifice.

There ought to be great cries in the land, great anger. But there is a strange silence. Why? Because we are all implicated. We have drifted to this dark unacceptable place together. We took the success of our economy as proof of the rightness of its underlying philosophy. We are now at a crossroad. Our future depends not on whether we get through this, but on how deeply and truthfully we examine its causes ... What we need now more than ever is a vision beyond the event, a vision of renewal.

As one looks over the landscape of contemporary events, one thing becomes very striking. The people to whom we have delegated decision-making in economic matters cannot be unaware of the consequences. Those whose decisions have led to the economic collapse reveal to us how profoundly lacking in vision they were. This is not surprising. These were never people of vision. They are capable of making decisions in the economic sphere, but how these decisions relate to the wider world was never part of their mental make-up. This is a great flaw of our world.

To whom do we turn for guidance in our modern world? Teachers have had their scope limited by the prevailing fashions of education. Artists have become more appreciated for scandal than for important revelations about our lives. Writers are entertainers, provocateurs or, if truly serious — more or less ignored. The Church speaks with a broken voice. Politicians are more guided by polls than by vision. We have disembowelled our oracles. Anybody who claims to have something to say is immediately suspect.

So now that we have taken a blowtorch to the idea of sages, guides, bards, holy fools, seers, what is left in our cultural landscape? Scientific rationality has proved inadequate to the unpredictabilities of the times ... This is where we step out into a new space. What is most missing in the landscape of our times is the sustaining power of myths that we can live by.

If we need a new vision for our times, what might it be? A vision that arises from necessity or one that orientates us towards a new future? I favour the latter.[7] It is too late to react only from necessity. One of our much neglected qualities is our creative ability to reshape our world. Our planet is under threat. We need a new one-planet thinking.

We must bring back into society a deeper sense of the purpose of living. The unhappiness in so many lives ought to tell us that success alone is not enough. Material success has brought us to a strange spiritual and moral bankruptcy.

If we look at alcoholism rates, suicide rates and our sensation addiction, we must conclude that this banishment of higher things from the garden has not been a success. The more the society has succeeded, the more its heart has failed.

Everywhere parents are puzzled as to what to do with their children. Everywhere the children are puzzled as to what to do with themselves. The question everywhere is — you get your success and then what?

We need a new social consciousness. The poor and the hungry need to be the focus of our economic and social responsibility.

Every society has a legend about a treasure that is lost. The message of the Fisher King is as true now as ever. Find the grail that was lost. Find the values that were so crucial to the birth of our civilization, but were lost in the intoxication of its triumphs.

We can enter a new future only by reconnecting what is best in us, and adapting it to our times. Education ought to be more global; we need to restore the pre-eminence of character over show, and wisdom over cleverness. We need to be more a people of the world.

All great cultures renew themselves by accepting the challenges of their times, and like the biblical David, forge their vision and courage in the secret laboratory of the wild, wrestling with their demons, and perfecting their character. We must transform ourselves or perish.

Reprinted from *The Times*, October 30, 2008, and *S&MN Review* no. 98. Ben Okri is the prize-winning Nigerian poet and novelist.

Endnotes

Note: See Bibliography for full references.
ISIS = Institute of Science in Society
SiS = Science in Society magazine
S&MN = Scientific and Medical Network

Introduction

1. The Scientific and Medical Network is a leading international forum for people engaged in creating a new worldview for the twenty-first century, bringing together scientists, doctors, psychologists, engineers, philosophers, complementary practitioners and other professionals. Formed in 1973, it has members in more than thirty countries.
2. Viktor Schauberger possessed this rare gift. He noted: 'The majority believes that everything hard to comprehend must be very profound. This is incorrect. What is hard to understand is what is immature, unclear and often false. The highest wisdom is simple and passes through the brain directly into the heart.'
3. Chris Clarke, 'The Implications of Modern Science for a New World View,' S&MN *Review*, no. 71.
4. The Law of Conservation of Energy states that the amount of energy throughout the universe is finite; energy merely transfers from one form to another — potential to kinetic and vice versa (the physical sphere). The Law of Anti-Conservation of Energy Law postulated by Viktor Schauberger holds that the amount of available energy — potential, kinetic or dynamic — can be increased at will to virtually any order of magnitude (the quantum sphere). Schauberger saw the two as dialectic counterparts.
5. H.H. Price, Wykeham Professor of Logic at Oxford University, writes: 'We must conclude, I think, that there is no room for telepathy in a materialistic universe. Telepathy is something which ought not to happen at all, if the materialistic theory were true. But it does happen. So there must be something seriously wrong with the materialistic theory, however numerous and imposing the normal facts which support it may be.' (*Hibbert Journal*, 1949) Goethe, too, said of conventional scientists: 'Whatever you cannot calculate, you do not think is real.'
6. Ho, 'Medicine in a New Key', ISIS.
7. Mae-Wan Ho, 'Medicine in a New Key,' ISIS.
8. Fritjof Capra, 'The *Yin Yang* Balance,' *Resurgence*, May 1981.
9. 'Quantum theory reveals a basic oneness of the universe ... As we penetrate into matter, Nature does not show us any isolated "basic building blocks", but rather appears as a complicated web of relations between the various parts of the whole ... The human observer constitutes the final link in the chain of observational processes, and the properties of any atomic object can be understood only in terms of the object's interaction with the observer.' Fritjof Capra, *The Tao of Physics*.
   Eastern mysticism shares similar concepts: 'The material object becomes ... something different from what we now see, not a separate object in the background or in the environment of the rest of Nature, but an indivisible part, and even in a subtle way an expression of the unity of all that we see.' Sri Aurobindo, *The Synthesis of Yoga*.

Chapter 1

1. This describes water perfectly: the ground is our base, where we come from, our common denominator; our Being is our very nature, our true integrity, our wholeness. Goethe was a scientist and polymath, as well as philosopher and poet.
2. 'The Ox and the Chamois,' in *Tau* magazine, no.146, p. 30.
3. A BBC film crew shot a remarkable film of a surfer inside a four-metre barrel wave for a Natural History Unit South Pacific series shown in 2009. Filmed in super slow motion using a high-definition camera, it shows the wave forming recognizable multiple fractal-like vortices shooting back from the face of the wave.
4. The vortex is like a door between dimensions. Black holes are vortices, connecting universes. One is reminded of what people describe who have had a near-death experience — going through a tunnel to meet their loved ones on the higher plane.
5. Mae-Wan Ho: 'O_2 dropping faster than CO_2 rising.' New research shows oxygen depletion in the atmosphere accelerating since 2003; bad news for mammals. (ISIS, 19 August 2009.)
6. The research of Viktor Schauberger, of whom more later.
7. Enzymes are proteins which increase the efficiency of biological cells.
8. In 1913 the chemist and natural theologian Lawrence J. Henderson pointed out that the strangeness of water consists in its possession of the precise properties that make it 'fit' for life on Earth. See *The Fitness of the Environment.*
9. See *The Water Wizard*, by Viktor Schauberger.
10. The Phoenicians called water *mem*, the root for 'memory', a reminder of the ancient belief in water's ability to record and transfer information. *Mem* is also the thirteenth letter in the Hebrew alphabet; their word for water is *mayim.*

Chapter 2

1. The postulation of a Goldilocks Zone has its limitations. For one thing, even on our own planet, micro-organisms like bacteria or microbes can live or even thrive in conditions that are prohibitive to most of life on Earth. These are called extremophiles: for instance, an organism that can thrive at temperatures between 80°–121°C, such as those found near volcanic vents on deep ocean floors, or the organisms adapted to the very high oxygen levels of the sub-glacial Lake Vostok in Antarctica.
   It is possible, even likely, that forms of proto-life are much commoner than we think — bacterial life that existed on Earth two billion years ago (and still does today) may well exist in environments like that of Saturn's moon Enceladus.
2. The most recent mission to search for planets like our Earth is the Kepler probe which NASA launched on 7 March 2009 from Cape Canaveral. This $3^1/_2$-year mission will search our own galaxy for Earth-sized planets in habitable zones around stars. It would take more sophisticated future missions to analyse their atmospheres to detect whether they could support life.
3. An Astronomical Unit (AU) is the distance from Earth to the Sun — approximately 93 million miles.
4. John McCreary in *The American Dowser*, November 1981.
5. Rob Gourlay, Australian groundwater specialist at www.eric.co.au.
6. *The Guardian,* 10 March, 2006
7. The idea of Earth having a 'consciousness' as part of its evolution is a variation of the 'Gaia hypothesis'.
8. Julian Caldecott, *Water, the Causes, Costs and Future of a Global Crisis.*
9. *When the Earth Nearly Died,* by D.S. Allan and J.B. Delair.
10. See Chapter 16, note 17.
11. 'Oceans and Global Warming,' *SiS*, 21 July, 2006.

Chapter 3

1. This quote is from the informative website of Prof. Martin Chaplin, biochemist at London South Bank University, to whom I am also grateful for permission to reprint models of the geometry of water structures from www.isbu.ac.uk/water. He identifies 67 anomalies of water: 12 of phase (states controlled by temperature or pressure), 20 of density, 12 material, 11 thermodynamic (for instance, specific heat) and 12 physical (such as viscosity). A summary of the less complex is given in Appendix 2.
2. Electromagnetic fields are produced by living organisms, for instance, the electric currents that flow in nerves and muscles.
3. Paolo Consigli, *Water Pure and Simple.*
4. A sharing of pairs of electrons between atoms.
5. Specific heat of other substances: ethyl alcohol 0.54, wood 0.42 , aluminium 0.21, iron 0.12, glass 0.11, copper 0.9, silver 0.6, gold 0.3.
6. There have been breakdowns in climatic self-regulation, which precipitated massive climate change and species' extinctions (for instance, during the Permian period, 250 million years ago), but then perhaps another self-regulating system at a higher level kicked in, to allow an evolutionary jump to a higher level of complexity to take place. However, it may be that this is more connected with Earth's evolution, which will anyway stimulate Nature's evolution.
7. *SiS* no. 15.
8. See *The Rainbow and the Worm*, by Mae-Wan Ho.
9. Listed on Martin Chaplin's website: www.isbu.ac.uk/water

Chapter 4

1. *Implosion* magazine, no. 8, 1945.
2. *Earth*, by Iain Stewart and John Lynch.
3. Vananda Shiva, *Resurgence*, November 2009.
4. *Hutchinson Encyclopedia.*
5. For instance, the waters of the Sahara.
6. Anything which obstructs natural replenishment can quickly exhaust over-exploitation, for instance,. land development and building, roads and car parks, inappropriate crops (cotton) or trees (thirsty eucalyptus); when swamps and wetlands are drained for farming; when the flow of water is accelerated, by straightening rivers, preventing floods, all prevent water from sinking into the ground.

When an aquifer under a city is drained, land subsidence follows (7.5 metres in Mexico City in the last 100 years). When commercial plantations and urban fill-in replace natural ecosystems and small farms, flash floods will increase and urban pollution will contaminate the surface and ground-waters.

Counties importing amounts in excess of their own supplies: Libya 711%, Saudi Arabia 722% United Arab Emirates 1,553%, which does not include importation of bottled drinking water: for instance, litres used annually per person: France 145, Spain 137, Mexico 169, Italy 184 (ref: Caldecott).

7. Véronique Mistiaen, 'Guarding Russia's sacred sea,' *Guardian Weekly,* 25 April, 2008.

Chapter 5

1. The magnetic poles are not fixed, but tend to shift in the course of a year. Quite recently their strength has shown a marked decrease, which may indicate that the north and south magnetic poles might soon change places. This gradual weakening, with magnetic eddies developing all over the planet until they concentrate at the opposite pole, occurs regularly every 500,000 years or so, and we are overdue for one. The effects would be a temporary loss of the magnetic shield that protects us from solar storms, occasional events that have in the past knocked out electricity networks and communication satellites. It would also, until the new poles were

reestablished, result in worldwide displays of the aurorae. (Iain Stewart, *Earth, the Power of the Planet.*)
2. *Nexus*, April 2009.
3. In eastern Canada these can create major power failures.
4. However, noctilucent clouds can form at about 80 km altitude in high latitudes in summer where meteoric dust particles act as nuclei for ice crystals when traces of water vapour are carried upwards by high-level convection caused by the vertical decrease of temperature in the mesosphere. Noctilucent clouds often have a shimmering, opalescent quality. As seen from the space shuttle, the aurorae form concentric ionising rings around the magnetic poles.
5. *Wikipedia.*

Chapter 6

1. Note: The temperatures shown are hypothetical, only to demonstrate the process.
2. The Australian dowser Alanna Moore describes a group pilgrimage down the Wimmera River in Victoria, teaching river care to the whole community, in her book *The Wisdom of Water.*
3. Julian Caldecott, *Water, the Causes, Costs and Future of a Global Crisis.*
4. Callum Coats in *Living Energies,* pp.176–7 describes one he studied.

Chapter 7

1. *Biological water:* The evolution of water structures through the geological ages is demonstrated by the complexity of water-forms found in human biological water today. There are approximately thirty forms, all of which are purpose-specific drivers of metabolic and physical processes in the body, all working in harmony:
   Digestive — bile, chyle, chyme, mucus, saliva, vomit, feces; *Intra-cellular* — drives cellular metabolic processes; *Extra-cellular* — provides cells with nutrients and waste removal, lymph; *Superficial* — sweat, sebum, tears; *Blood* — plasma, serum, pus and urine; *Cerebrospinal fluid* — surrounding brain and spinal cord; *Sexual and Reproductive* — includes menses, amniotic fluid and breast milk; *Synovial fluid* — surrounding bone joints; *Miscellaneous* — pleural fluid, cerumen (earwax), aqueous humour (eyes). *Source: Wikipedia*
2. Karol Sikora, *The Observer*, 18 May, 2008.
3. BBC Radio 4, 'Today' Programme, 23 June, 2008.
4. *The Golden Fountain* by Coen Van der Kroon.

Chapter 8

1. Schauberger identified subtle energies of the fourth and fifth dimensions (see Chapter 10) as responsible for this alchemical process. He named them: *dynagens* as the primal male (Sun) energies, which initiate growth; *fructigens* the feminine (Earth) energies which symbolize fruitfulness; and *qualigens* which determine the quality of life.
2. Viktor Schauberger, 'The Dying Forest' *(Der sterbende Wald)* Part 1, *Tau* magazine, Vol. 151, November 1936, p. 30.
3. Writing 2,300 years ago, in his *Critias,* Plato described how Attica's mountains a century or two before had been covered in verdant forest, and her fertile plains had deep soil, that by his day had become stony shingle. The rainfall and the soil had disappeared because the forests had been cut down.
4. BBC 'Today' Programme, 13 October, 2008.
5. BBC 'Today' Programme, interview with Jim Naughtie, 15 May, 2008.
6. www.orangutan.org.uk
7. James Lovelock is also pessimistic. He believes recent deforestation to be the cause of global warming, because Earth's forests are the principal regulator of climate. In his book The Vanishing Face of Gaia, he puts the case that global warming is now unstoppable, and that the human population that Earth was able to sustain by the end of this century could be as low as 1 billion people. They

would be limited to those environments remaining habitable, such as Northern Europe, Siberia, Canada, Japan and southern South America."
Lovelock comments (Observer, 22 Mar '09): Over the last million years, several climatic events brought decimation of numbers of the human species, yet each trauma seemed to herald an evolutionary advance; eg between the ice ages, sea level rose 120 metres, flooding the plains, but Homo Sapiens emerged (see Chap 11, Chaos introduces higher energies). He believes that humanity should benefit from the coming population collapse."

8. The north-west coasts of North America, the south-west of South America, small segments in north-west Europe, Japan, south-east Australia and the west coast of South Island, New Zealand.
9. David Attenborough's TV documentary on the Pacific salmon in 'Nature's Great Events': BBC, 18 February, 2009.
10. Peter Bunyard, *The Breakdown of Climate.*

Chapter 9

1. Carl Safina, 'For Evolution to Live, Darwin Must Die,' Observer, 8 March, 2009.
2. Charles Darwin, *The Origin of Species.*
3. Mae-Wan Ho, 'Death of the Central Dogma,' *Science in Society,* 3 September, 2004.
4. See Darwin's 1886 letter to M. Wagner, in *Charles Darwin: Life and Letters*, ed. F. Darwin.
5. Mae-Wan Ho identifies the antiquated and destructive worldview that dominates medicine, biology, economic and social/political systems with the contraction 'Domo', meaning 'dominating model'; the powerful, controlling and seductive anti-life tool of the world seen as machine.
6. See: www.emofree.com.
7. Bruce Lipton, *The Biology of Belief.* This insightful introduction to the 'New Biology' is inspired by epigenetic theory. Bruce Lipton, an award-winning cell biologist who taught in medical schools for twenty years, has empowered thousands of people with insights to make profound changes in their lives with his popular workshops and lectures.
8. The comparison with a birth chart is apposite. Many (for instance, the tabloid newspapers) regard your sun sign (and other aspects of your chart) deterministically, as 'old biology' does the DNA. To the serious astrologer, however, the birth chart is a blueprint that can be transcended, in the same way that epigenetics shows how the genetic templates of the DNA can be transcended.
9. Such as the effects of harmful medication which cause iatrogenic illness (or the 'possible side effects' listed on the leaflet that comes with your pills), now the main source of death in the USA (>300,000 per year: see Bruce Lipton's study of US Government 2003 statistics).
10. Mae-Wan Ho, 'Epigenetic Inheritance — What Genes Remember,' *SiS*, 41, 2009.
11. Marcus Pembrey, 'Sins of the fathers, and their fathers,' *European Journal of Human Genetics,* 2006.
12. *Genesis of the Universe, The Ancient Science of Continuous Creation*, by Paul LaViolette.
13. Mae-Wan Ho, 'Quantum Jazz, the Tao of Biology.'
14. See also: *The Tao of Physics* by Fritjof Capra.

Chapter 10

1. Requirements for living water. See the principal qualities essential for living water in Appendix 1: Water and Health.
2. The extra 'e' in 'carbone' enlarges the usual range of elements used in forming the physical structures of life, except oxygen and hydrogen. (Schauberger also classified the elements as masculine or feminine.)

Chapter 11

1. Theosophy is a metaphysical philosophy originating with the Russian philosopher Helena Petrovna Blavatsky in 1875. Theosophists trace its ancient origin to the universal striving for spiritual knowledge that has existed in all cultures. They believe that all religions are attempts by spiritual 'Masters' to help humanity evolve to a higher level of consciousness. It has strong links to esoteric Buddhism.
2. *Dogs That Know when their Owners are Coming Home,* by Rupert Sheldrake.
3. See above Chapter 3, and Mae-Wan Ho, 'Two States Water Explains All?' (*SiS*, 25 Oct '06).
4. *The Rainbow and the Worm* by Mae-Wan Ho.
5. Dr Mae-Wan Ho, who has a world reputation in the new science of the organism, resigned from the Open University, refusing to condone the growing funding of research in genetics by biotechnical companies. Recognizing the need for transparency and open debate on public issues, in 1999 she founded with her husband, Prof. Peter Saunders, the Institute of Science in Society. Besides its productive conference and research programme, it also publishes a bi-monthly magazine.

Mae-Wan Ho defies pigeon-holing — she is a genuine polymath — and is an outspoken and tireless campaigner who challenges the bastions of power in the scientific establishment (see Links).

6. Do read Mae-Wan Ho's background article 'Quantum Jazz — the Tao of Life' on www.i-sis.org.uk. The science behind this research is contained in her remarkable book *The Rainbow and the Worm, the Physics of Organisms*, more popular presentations of which are her 'New Age of Water' series of articles found on the Institute's website (as above).
7. 'Quantum Coherent Liquid Crystalline Organism,' Energy Medicine Conference, Copenhagen, 19 September 2008.
8. Ibid.
9. Ho's simple diagram of the conservation of energy in organisms is also a model for the successful working of any organization. If this were applied to the nation's financial policy we would not have seen the appalling haemorrhaging of money and resources which have resulted in entropy and waste. It could also transform the way our communities are run, and our energy and food is produced. It is the key to a sustainable human society, and to living lightly on the Earth.

   Ho has published a very telling example of new concepts of food production which depend on the recycling of energy from one part of a model farm to another (diverse crops and animals, composting, aquaculture, worms, mushrooms, biogas and hydrogen production, and so on). See 'Dream Farm,' *SiS,* iss.38.
10. 'Quantum Jazz, the Tao of Biology.' See: www.i-sis.org.uk
11. Zheng, J-M, and Pollack, G.H. 'Long-range forces extending from polymer-gel surfaces'. *Physical Review,* E 2003, 68, 031408. See also M-W. Ho, 'Water forms massive exclusion zones', New Age of Water series, ISIS, 2004.
12. ISIS, 31 March, 2008.
13. One of the 'kitchen table' experiments of Gerald Pollack, Professor of Bio-engineering, University of Washington, Seattle.
14. 'Quantum Coherence and Conscious Experience,' by Mae-Wan Ho, *Kybernetes* 26, pp. 263–76, 1997.
15. Mae-Wan Ho, 'Collagen Structure Revealed', New Age of Water Series, ISIS, 23 October, 2006.
16. Fullerton, G.D. and Amurao, M.R. 'Evidence that collagen and tendon have monolayer water coverage in the native state', *Int. J. Cell Biol.* 2006, 30, pp. 56–65.
17. Mae-Wan Ho, 'Coherent Energy, Liquid Crystallinity and Acupuncture': talk to British Acupuncture Society, 2 October, 1999.
18. *Op. cit.*

Chapter 12

1. Ilya Prigogine's pioneering research in self-organizing systems, based on his dissipative systems theory, became a cornerstone of quantum physics.
2. Callum Coats gives a helpful explanation of dialectic thinking, which he insists is imperative for comprehension of the whole *(Living Energies*, p. 63). Thus, *yin* is balanced by *yang*, magnetism by electricism, frequency by wavelength, spirit by matter, and so on. Thus also, *Chaos x Order = 1*. Without chaos (undifferentiated, unstructured matter or energy, or unordered, unmetamorphosed unconditional love) there could be no basis for the creation of order (differentiated, harmonically-structured matter or energy; therefore the foundation for order is chaos.
3. One such candidate might be the Transition movement (see Chapter 19), which seems to have a 'fractal' way of multiplying. (Also see Appendix 3: Ben Okri on 'The Bankruptcy of our Civilization'.)
4. James Clerk Maxwell, the founder of electromagnetic theory, by challenging the omnipotence of the Second Law of Thermodynamics in 1871 with his 'Maxwell's Demon' proposition, anticipated the basis of quantum physics.
5. Callum Coats clarifies the relationship between phi and the egg shape in *Living Energies,* pp. 65–72. The numerical value of *phi* is 1:1.6187, or 5:8 (a $\sqrt{3}$ rectangle).
6. This is why you should always store your eggs pointy end up.
7. See: www.Sulis-Health.co.uk
8. Schauberger believed that a water body (for instance, a stream) can communicate with our own water body (see above Chapter 10). Cleve Backster's research suggests this might even happen at a distance (see below Chapter 14).

Chapter 13

1. George Gurdjieff, the Caucasian mystic, said that our brains are principal receivers of cosmic energy.
2. Schwenk, *Sensitive Chaos*, p. 68.
3. Lawrence Edwards' *The Vortex of Life* is the fascinating record of his rigorous research. He shows how the Moon, particularly when amplified by certain aspects of Saturn and Mars, can create a cyclical effect in the growth of tree buds.
4. August Schmauss, 'Biologische Gedanken in der Meteorologie,' *Forschungen und Fortschritte*, Vol. 21, 19.
5. Paul Raethjen, *Dynamics of Cyclones,* Leipzig, 1953.

Chapter 14

1. An exception was Albert Einstein who was known to have experimented with dowsing (Alanna Moore, *The Wisdom of Water*).
2. *The Journal of Scientific Exploration*, Stanford University, 1995.
3. Cleve Backster, *Primary Perception*, p. 34.
4. Elisabeth Sahtouris: *Earth Dance — Living Systems in Evolution*.
5. *Primary Perception*, p. 41.
6. Backster's controversial research was first published in 1973 in *The Secret Life of Plants*, by Peter Tompkins and Christopher Bird. It took another thirty years for Backster to publish his own account of his fascinating research in his book *Primary Perception* which was well received by his peers: 'Cleve Backster's research has profound implications for humanity and its future evolution.' *Deepak Chopra*. 'The implications of this work are enormous for science, society and ecology.' *Jean Houston*.
7. Chris Clarke, 'Entanglement — the Explanation for Everything?' S&MN *Review*, no. 86.

Chapter 15

1. Monika Fuxreiter's research at the Institute of Embryology, Budapest

(cited by Mae-Wan Ho) demonstrates that DNA is inseparable from water. It may be that water plays the role of recording the information to be conveyed to the DNA.
2. M-W. Ho, 'The Strangeness of Water and Homeopathic "Memory".' ISIS, 31 May 2002
3. M-W. Ho, 'Crystal Clear — Messages from Water': ISIS, 1 June 2002
4. *The Secret Power of Music.*
5. *Your Body Doesn't Lie..*
6. *Health and Light*, by John Ott.
7. *cf.* Quantum Jazz, p. 103.
8. My mother believed in the inherent goodness in all people. That this was not just naiveté was shown by the fact that some people with a poor social reputation could respond to her with unaccustomed cooperation. However, in the past fifty years, basic standards of morality have given way to greater self-indulgence and lack of honesty.
9. See Epigenetics, Chapter 9.

Chapter 16
1. Is it accidental that the largest concentration of megalithic monuments of spiritual significance in Britain are found on the Orkney Islands, the only area in the British Isles where mineable uranium can be found?
2. *National Geographic* special edition on Water, 1993.
3. Prof. Jeffrey Sachs, UN Millennium Project, New Delhi, January, 2007.
4. *The Guardian*, 12 May 2008.
5. *Guardian Weekly*, 13 February 2009.
6. *The Ecologist*, May 1999.
7. International Management Institute.
8. *Guardian Weekly*, March 2001.
9. *National Geographic*, 'Earth's Fresh Water under Pressure,' September 2002.
10. *National Geographic*, ibid.
11. *The Ecologist*, May 1999.
12. *The Ecologist*, May 1999.
13. *Ibid.* Caspar Henderson.
14. *Ibid.*
15. See also Appendix 1, Water & Health.
16. Viktor Schauberger, *Nature as Teacher.*
17. *Fluoride: Drinking ourselves to Death?* by Barry Groves, a well informed source on this topic.
18. Waldblott, McKinney and Burgstahler: *Fluoridation — The Great Dilemma*, p. 288.
19. *Journal of Dental Research*, 1990; 69, pp. 723–7.
20. In Australia, some of the fluoride laws are so draconian that people may be prosecuted for speaking out against water fluoridation. 'Living in a democratic fluoridated country,' *Australian Fluoridation News.*
21. Barry Groves, *Fluoride: Drinking Ourselves to Death?* p. 227.
22. Viktor Schauberger, *Our Senseless Toil,* Part II, p. 14.
23. Essentially growth-promoting dynagens, created by the bi-metal composition: silver (male) and copper (female). The silver also has bactericidal properties. Dynagens are also produced by the centripetal movement of the main water body, raising the overall vitality, life-energy and wholesomeness of the water.
24. *The Wisdom of Water* by Alanna Moore.

Chapter 17
1. The Uru Chipaya, the ancient 'water people' of Bolivia blame their upstream neighbours for diverting precious supplies. (*Guardian Weekly*, 1 May, 2009.) Competition over the water of the River Jordan is a flashpoint in the Near East.
2. BBC Radio 4: 'Home Planet,' 24 March, 2009.
3. Bushnell D. 'Seawater/Saline Agriculture for Energy, Warming, Water, Rainfall, Land, Food and Minerals,' 2008. See web address: http://web.mac.com/savegaia/flowerswar/Project/Entr%C3%A9es/2008/12/5_mise_%C3%A0_jour_en_cours_files/Dennis-Bushnell-saline-agriculture.pdf

See also Glenn, E.P., Brown, J.J. and O'Leary, J.W., 'Irrigating crops with seawater,' *Scientific American,* August, 1998, pp. 76–81.

4. The Water Association (1992) is a pioneering European group of like-minded scientifically trained innovators of sustainable systems for purifying and re-energizing waste water and other water resources.
5. Alanna Moore, *The Wisdom of Water.*
6. *Guardian Weekly*, 17 July, 2009.
7. *The Guardian*, 19 May, 2009.
8. *Guardian Weekly*, 23 May, 2005.
9. Christopher Field of the Carnegie Institute of Science testified: 'With severe drought from California to Oklahoma, a broad swath of the south-west is basically robbed of having a sustainable lifestyle. We are close to a threshold in a very large number of American cities where uncomfortable heat waves make cities uninhabitable. Sacramento could face heatwaves for up to 100 days a year.' *Guardian Weekly*, 6 March, 2009.
10. *Guardian Weekly,* 26 June, 2009.
11. *Guardian Weekly,* 17 July, 2009.
12. It is more marked than in the Antarctic because most of the world's CO_2 emissions originate in the Northern Hemisphere.
13. See University of Hamburg: animated video of ice cover changes on Arctic Ocean over nine years: www.youtube.com/watch?v=e2rt1QWC-9Q
14. A German plan to build most advanced polar research vessel in the world is being funded by the European Commission, to be commissioned in 2012. The research icebreaker 'Aurora Borealis' will have a multi-functional role of deep-sea drilling and supporting climate/environmental research in both Arctic and Antarctic regions. See eri-aurora-borealis.eu
15. Crystalline methane hydrides are also deposited as on the continental shelves, especially off the east coasts of North and South America. They are stable within a limited temperature range, but it is believed they could be suddenly released if the sea temperature rose to a critical level. No one knows just what that tipping point might be, but, when it comes, the release of methane would be fast and catastrophic to global warming levels, as methane is twenty times more powerful as a greenhouse gas than CO_2.
16. See also geological time chart, Figure 1, Chapter 2 above.
17. Chris Bowler, marine biologist on the EU Tara Ocean project in Barcelona, setting out on a three-year survey of the world's oceans to study life forms.
18. *The Observer*, 4 October, 2009: Prof. J-P. Gattuso of the Centre National de la Recherche Scientifique, at a European Commission 'Oceans of Tomorrow' international conference in Barcelona, 6 October, 2009, reported research from Svalbard (Spitzbergen) projecting that of the Arctic Ocean will be 10% corrosively acidic by 2018; 50% by 2050 and 100% by 2100. This has the most serious implications for ocean biodiversity, for the bottom of the food chain, from small molluscs to the vast numbers of zooplankton, will be particularly vulnerable and would not survive such changes. The acid dissolves the bony structure of these tiny animals. Bio-engineering technology to solve this crisis was discussed at the conference, but quickly dismissed as impractical. Probably the only way to slow down this collapse is the immediate abandonment of fossil fuels and/or the precipitate reduction of the human population to a third of its present level. The Northern Hemisphere receives most human-produced CO_2, but we should expect this phenomenon to spread later to the Southern.
19. Both Viktor and Walter Schauberger criticized people who tried to produce 'free' energy with no understanding of what was involved. They claimed that the energies of the fifth and six dimensions required a state of humility and commitment in order to be cooperative. Viktor frequently insisted that the first priority was to observe and learn from Nature.
20. 'Can Water Burn?' ISIS, 10 June, 2009.

Chapter 18

1. Mae-Wan Ho, in her book *Genetic Engineering — Dream or Nightmare?* describes the large-scale release of transgenic organisms as 'much worse than nuclear weapons as a means of mass destruction — as genes can replicate indefinitely, spread and recombine'.
2. See 'Water in Australian Landscapes' in *The Wisdom of Water* by Alanna Moore.
3. See www.naturalsequenceassociation.org.au.
4. *Silent Spring* by Rachel Carson.
5. *Food Futures Now, Organic, Sustainable, Fossil Fuel Free,* Mae-Wan Ho, Sam Burcher, and L.C. Lim, ISIS.
6. The ancient round towers found in Ireland are thought to have been built to enhance the fertility of the soil by creating paramagnetic (yang) energy which originates in the Sun, whereas plants produce diamagnetic (*yin*) energy, stimulated by water (see *Stone Age Farming* by Alanna Moore).
7. The use of human 'nightsoil' used to be widespread. This practice is now confined to Himalayan communities.
8. 'Science of the Organism and Sustainable Systems: Implications for Agricultural Policies,' *SiS*, March 2009.
9. Rebecca Hosking, 'A Farm for the Future,' *Daily Mail*, 15 February, 2009, and BBC2 film.
10. The Transition Network was formed in 2003 in Totnes, Devon, to meet the twin challenges of Peak Oil and climate change, emphasizing the values of community strength and self-sufficiency. It has grown fast, and is now helping to transform hundreds of enthusiastic communities in many countries around the world. See web page: www.transitiontowns.org
11. *Food Futures Now, op.cit.*
12. Described in *Secrets of the Soil*, Peter Tompkins and Christopher Bird, 1998.

Chapter 19. Epilogue

1. James Lovelock, on a visit to the Hadley Centre, the UK's primary climatic research centre, while impressed by each scientist's expertise in his own area, noted that there was little overall recognition of the Earth as a dynamic system in critical imbalance. (Royal Society lecture, 18 October 2007.)
2. I find being close to horses a bit scary. The field where I walk regularly has three mares with their growing foals just now. I like to take them apples for a treat, but when all six surround you, they can feel a bit threatening. When I tried to visualize these lovely, powerful animals as part of my family through the water link, they seemed to become more like family and more friendly. This is what you might call a holistic connection. Try it yourself; it can even work with people!
3. In M-W. Ho's *The Rainbow and Worm.*
4. This has holistic parallels to Michaela Wright's 'Perilandra' system which identifies any creative project as a 'garden', a projection of her carefully cultivated bio-diverse garden in Virginia, which is over-lighted by 'spiritual guides'; or to the 1960's Findhorn Garden in the north of Scotland with its 'devas'.
5. See Ben Okri on 'The Moral Bankruptcy of our Civilization.' (Appendix 3).
6. www.transitiontowns.org.
7. See Wilkinson, Richard and Pickett, Kate, *The Spirit Level — Why more equal Societies almost always do Better.* Elsewhere Herman Daly asks: 'What is the proper range of inequality (the highest income as a multiple of the lowest) — one that rewards real differences and contributions rather than just multiplying privilege? Plato thought it was a factor of 4. Universities, civil servants and the military seem to manage with a factor of 10–20. In the US corporate sector it is 500–1000.' ('Towards a Steady-State Economy,' *Resurgence.*) Endless economic growth, irresponsible banking and advertising are basically unsustainable. A move away from these towards an awareness of fulfilling the

needs of society as a whole would reduce inequalities.

8. *The Information Age.* There is an old saying is that the flutter of a butterfly's wings on the far side of the world can affect the destiny of nations. A small drama — Governor Sarah Palin's indiscretion — in the remote state of Alaska in the 2008 US presidential election campaign, may have affected the future of the whole world community. The reaction of Far Eastern financial markets to a political decision in one European country, instantaneously affects world economic health. Individual countries are now powerless to determine their own futures. National boundaries are meaningless in the new information age. The Internet has done more to link together the world community than any other influence. It is less easily controlled than television for political and commercial purposes, and can therefore be a powerful medium for creative change.
9. David Korten, *The Great Turning: From Empire to Earth Community.*
10. See Coats, *Living Energies*, p. 63.
11. See *The Field* by Lynne McTaggart.
12. Callum Coats describes the correspondences as dialectic pairs, which together make a whole (*Living Energies*, p. 63). John baptized Jesus with Water and the Spirit, a metaphor for the Water Domain and the Etheric Field, which together are Unity.

Appendices

1. Viktor Schauberger gave the term 'carbone' to the loosely named 'carbonous' (more than carbon) building blocks of matter. The exceptions were oxygen and hydrogen.
2. The rule that city tap water must not contain E.coli or fecal coliform bacteria does not apply to bottled water, nor is it required to be filtered or disinfected as is city water. City tap water is required to meet standards for certain toxic or cancer-causing chemicals, like phthalate (a chemical that can leach from plastic bottles); the industry persuaded the Federal Drugs Agency to exempt bottled water from these requirements. Water supply companies must tell consumers what is in their water. The bottling industry successfully killed this 'right to know' requirement for bottled water.
3. Twice the size of Texas, this enormous garbage dump, composed mostly of small fragments of plastic, is formed by the clockwise circulating currents of the North Pacific Gyre and is a great hazard to fish.
4. According to the U.S. Environmental Protection Agency (EPA): 'Every home in America has an elevated level of chloroform gas present due to the vaporization of chlorine ... from tap and shower water.' Chlorine vapours are an irritant which can cause respiratory problems such as asthma, bronchitis, and allergies. Point-of-entry, whole house filtration results in the effective reduction of chlorine, VOCs (volatile organic compounds) and other chemicals that vaporize (for instance, from clothes in the washing machine) and contaminate the indoor air.
5. You will find them at www.Sulis-Health.co.uk
6. See Epilogue.
7. See Epilogue, 'Holism in Society'.

References and Further Reading

Note: S&MN = Scientific and Medical Network; ISIS = Institute of Science in Society; *SiS* = Science in Society magazine

Allan, D.S. & Delair J.B., *When the Earth Nearly Died,* Gateway, Bath 1992. *(The cause of the Great Flood?)*

—, *Catastrophe,* Bear & Co, Rochester VT, 1999. *(US edition of the above.)*

Alexandersson, Olof, *Living Water — Viktor Schauberger and the Secrets of Natural Energy,* Turnstone, London, 1982.

Altman, Nathaniel, *Sacred Water — The Spiritual Source of Life,* Hidden Spring, 2002.

Ash, David & Hewitt, Peter, *The Vortex — Key to Future Science,* Gateway, Bath, 1983.

Backster, Cleve, *Primary Perception, Biocommunication with Plants, Living Foods and Human Cells,* White Rose Millennium Press, Anza CA, 2003. *(A classic.)*

Baker, Richard St Barbe, *I Planted Trees,* London, Lutterworth, 1944.

Ball, Philip, *Life's Matrix — A Biography of Water,* University of California, Berkeley, 2001.

Barry, R.G & Chorley, R.J., *Atmosphere, Weather & Climate,* Methuen, London, 1976.

Bartholomew, Alick, *Hidden Nature — the Startling Insights of Viktor Schauberger,* Floris Books, Edinburgh, 2003; rev.ed. 2006.

—, *The Schauberger Keys,* Schauberger, Bath, 2003. *(Notes on Schauberger's worldview.)*

—, 'How Nature Works,' *Living Lightly,* 17, 2001.

—, 'Nature is Sacred,' *Resurgence,* no. 225, 2004.

—, 'Towards a Science of Nature,' S&MN *Review,* no. 84.

—, 'The Evolution of Earth & of Life,' S&MN *Review,* no. 92.

—, 'Viktor Schauberger' in *Twentieth Century Visionaries,* Green Books, Totnes, 2007.

— 'What is Living Water?' *Caduceus,* 2007.

Bass, Karen (ed), BBC Natural History Unit, *Nature's Great Events — The Most Spectacular Natural Events on the Planet,* Beazley, London, 2009; University of Chicago, Chicago, 2009. *(To accompany BBC 6-part TV series, introduced by David Attenborough.)*

Batmanghelidj, Fereydoon, *Your Body's Many Cries for Water,* Global Health Solutions, 1995; *UK rev.ed,* Tagman, Norwich, 1997. *(Good.)*

—, *Water & Salt, Your Healer from Within,* Tagman, Norwich, 2003.

Bennett, J.G., (ed. A.G.E. Blake), *Deeper Man,* London, Turnstone, 1978. *(An exponent of the teachings of G.I. Gurdjieff, for instance, The Law of Three.)*

Bortoft, Henri, *Goethe's Science of Nature,* Floris Books, Edinburgh, 2002. *(Recommended.)*

Boulter, Michael, *Extinction, Evolution and the End of Man,* Fourth Estate, London, 2002.

Brennan, Barbara Ann, *Hands of Light — A Guide to Healing Through the Human Energy Field,* Bantam, London 1987.

Bruges, James, *The Big Earth Book —*

Ideas & Solutions for a Planet in Crisis, Sawday, Bristol, 2007.

Bunnett, R.B., *Physical Geography in Diagrams,* Longman, London, 1965.

Bunyard, Peter, *The Breakdown of Climate — Human Choices or Global Disaster?* Floris Books, Edinburgh, 1999.

—, (ed), *Gaia in Action: Science of the Living Earth,* Floris Books, Edinburgh, 1996.

— 'The Real Importance of the Amazon Rain Forest,' ISIS, Mar. 2010.

Button, John (ed), *The Best of Resurgence,* 25 years' selection, Resurgence, Bideford, 1991.

Caldecott, Julian, *Water: Life in Every Drop,* Transworld, London, 2007.

Capra, Fritjof, *The Tao of Physics,* London, Wildwood, 1975.

—, 'The Yin Yang Balance,' *Resurgence,* May 1981.

Carson, Rachel, *Silent Spring,* Houghton Mifflin, Boston, 1963. *(A classic.)*

—, *The Sea Around Us,* Harper, New York, 1951. *(Poetic.)*

Chaplin, Martin, 'The importance of cell water,' *SiS no.* 24, 42–45, 2004.

—, 'Water; its importance to life,' *Biochem. Mol. Biol. Educ.* 29 (2), 54–59, 2001.

Clarke, Chris, 'Entanglement – the Explanation for Everything,' S&MN *Review,* no. 86.

— 'The Implications of Modern Science for a New World View,' S&MN *Review,* no. 71.

Coats, Callum, *Living Energies – An Exposition of Concepts Related to the Theories of Viktor Schauberger,* Gateway, Bath, 1996. (*Most authoritative work on Schauberger's research.)*

Cobbald, Jane, *Viktor Schauberger — A Life of Learning from Nature,* Floris Books, Edinburgh, 2005. *(Recommended.)*

Cloos, Walther, *The Living Earth — The Organic Origin of Rocks and Minerals,* Lanthorn, UK, 1977. *(A Steiner earth science book.)*

Consigli, Paolo, *Water, pure and simple,* Watkins, London, 2008. *(Comprehensive and fascinating.)*

Cook, David, *The Natural Step — Towards a Sustainable Society,* Schumacher Briefing No. 11, Green Books, Totnes, 2004.

Crawford, E. A., *The Lunar Garden — Planting by the Moon Phases,* Weidenfeld & Nicholson, London, 1989.

Darwin, Charles, *The Origin of Species.* In many editions, the first being *The Origin of Species by Means of Natural Selection, or the Preservation of Favoured Races in the Struggle for Life,* London, Murray, 1858.

Darwin, F. (ed), *Charles Darwin: Life and Letters,* John Murray, London 1888.

Diamond, Jared, *Collapse — How Societies Choose to Fail or Succeed,* Viking, New York, 2005.

Diamond, John, *Your Body Doesn't Lie,* Harper & Row, New York, 1979.

Edwards, Lawrence, *The Vortex of Life — Nature's Patterns in Time & Space,* Floris Books, Edinburgh, 1993. *(Important research on biological planetary influence.)*

Emoto, Masaru, *The Hidden Messages in Water,* Beyond Words, Hillborough, 2001.

—, *The Secret Life of Water,* Beyond Words, Hillborough, 2006.

—, *The True Power of Water,* Beyond Words, Hillborough, 2005.

Endres, Klaus-Peter & Schad, Wolfgang, *Moon Rhythms in Nature,* Floris Books, Edinburgh, 2002.

Flannery, Tim, *The Weather Makers,* Grove, New York, 2005.

Forward, William & Wolpert, Andrew (eds.), *Chaos, Rhythm and Flow in Nature,* (The Golden Blade no. 6), Floris Books, Edinburgh, 1993.

Goodwin, Brian, *Nature's Due — Healing our Fragmented Culture,* Floris Books, Edinburgh, 2007.

Gordon-Brown, Ian, with Barbara Somers (edited by Hazel Marshall), *The Raincloud of Knowable Things,*

A Practical Guide of Transpersonal Psychology (workshops & method), Archive, Dorset, 2008.

x Graves, Tom, *Needles of Stone*, Turnstone, London, 1978 *(A dowsing classic.)*

Hageneder, Fred, *The Spirit of Trees — Science, Symbiosis, and Inspiration,* Floris Books, Edinburgh, 2005. *(A lovely book.)*

Hambling, Richard, *Clouds*, David & Charles, Newton Abbott, 2008.

Hamaker, John & Weaver, Don, *The Survival of Civilization,* Hamaker Weaver Publications, Burlingame CA, 1975.

Hanniford, Carla, *Smart Moves, Why learning is not all in the head,* Great Ocean Publications, 1995. *(Detailed research into the educational benefits of Brain Gym exercises.)*

Hall, Alan, *Water, Electricity and Health — Protecting Yourself from Electrostress at Home and Work,* Hawthorn, Stroud, 1997.

Harding, Stephan, *Animate Earth*, Green Books, Totnes, 2006.

x Henderson, Lawrence J., *The Fitness of the Environment [for life]*, Macmillan, New York, 1913.

Ho, Mae-Wan, *The Rainbow and the Worm — The Physics of Organisms,* World Scientific, Singapore, 1993. Rev. third edition 1998. *(A seminal work.)*

—, *Genetic Engineering — Dream or Nightmare?* Gateway, Bath, 1998; Continuum, New York, 2000.

—, 'Quantum Coherence and Conscious Experience,' *Kybernetes* 26, pp. 263–76, 1997.

—, 'Crystal Clear — Messages from Water.' New Age of Water Series, ISIS, 1 June, 2002.

—, 'The Strangeness of Water and Homeopathic "Memory",' New Age of Water Series, ISIS, 31 May , 2002.

—, 'Water forms massive exclusion zones,' New Age of Water series, ISIS, 2004.

—, 'Collagen Structure Revealed,' New Age of Water Series, ISIS, 23 October, 2006.

—, 'Dream Farm' *SiS,* 38, *also* 'Dream Farm 2, The Story So Far,' ISIS, 24 July, 2006.

—, 'Two States Water Explains All,' New Age of Water Series, *SiS*, 25 October, 2006.

—, 'Quantum Coherent Liquid Crystalline Organism,' Energy Medicine Conference, Copenhagen, 19 September, 2008.

—, 'Science of the Organism & Sustainable Systems — Implications for Agricultural Policies,' *SiS*, London, March, 2009.

—, 'O_2 dropping faster than CO_2 rising,' ISIS, 19 August, 2009.

—, 'Can Water Burn?' ISIS, 10 June, 2009.

—, & Burcher, Sam & Lim, L.C., *Food Futures Now — Organic, Sustainable, Fossil Fuel Free,* ISiS/TWN, London, 2008. *(Important.)*

Hollick, Malcolm, *The Science of Oneness — a World-view for the 21st Century,* O-Books, Ropley, Hants., 2006.

Hood, K., Halpern, C., Greenberg, G., & Lerner, R. (eds.), *Handbook of Developmental Science, Behavior & Genetics,* Blackwell, New York, 2009.

Hopkins, Rob, *The Transition Handbook — From Oil Dependence to Local Resilience*, Green Books, Totnes, 2008. *(Standard guide — important.)*

Hosking, Rebecca, 'A Farm for the Future,' *Daily Mail*, 15 February, 2009. *(Also an impressive BBC2 film, 19 February, 2009.)*

Kilgour, William, *Twenty Years on Ben Nevis* (Meteorological Station, 1882–1904), Gardner, Paisley, 1905.

Korten, David C., *When Corporations Rule the World*, Earthscan, London, 1996.

—, *The Great Turning – From Empire to Earth Community,* Berrett-Koehler, New York, 2006. (*Inspiring & prophetic.)*

—, *Agenda for a New Economy – From Phantom Wealth to Real Wealth,* Berrett-Koehler, New York, 2009. *(Why Wall Street can't be fixed, and how to replace it.)*

Kronberger, Hans & Lattacher, Siegbert, *On the Track of Water's Secret — from Viktor Schauberger to Johannes Grander,* Uranus, Vienna, 1995.

LaViolette, Paul, *Genesis of the Cosmos — The Ancient Science of Continuous Creation,* Bear, Rochester VT, 1995.

Lipton, Bruce, *The Biology of Belief — Unleashing the Power of Consciousness, Matter & Miracles,* Hay House, 2008. *(A seminal book.)*

Lockley, Martin, 'Intelligent Design Paradigm,' S&MN *Review,* no. 87.

Lorimer, David *et al* (eds.), *Wider Horizons: Explorations in Science and Human Experience,* Scientific & Medical Network, Leven, Fife, 1999.

Lovelock, James, The *Revenge of Gaia,* Allen Lane, London, 2006.

—, *The Vanishing Face of Gaia,* Basic Books, 2009.

Makarieva, A.M. & Gorshkov, V.G. 'Condensation-induced kinematics and dynamics of cyclones, hurricanes and tornadoes,' *Physics Letters A* 2009, 373, pp. 4201–5.

Manning, Jeane, *The Coming Energy Revolution — The Search for Free Energy,* Avery, New York, 1996.

Marks, William E., *The Holy Order of Water — Healing Earth's Waters and Ourselves,* Bell Pond, Great Barrington VT, 2001.

Marshall, Hazel, (ed), *The Wisdom of the Transpersonal,* a trilogy uniting the work of Barbara Somers & Ian Gordon-Brown *(q.v.)*

McTaggart, Lynne, *The Field — the Quest for the Secret Force of the Universe,* Harper Collins, London, 2003.

Merrifield, Jeff, *Damanhur — The Real Dream,* Thorsons, London, 1998. *(Account of an Italian artistic & spiritual community.)*

Moore, Alanna, *The Wisdom of Water,* Python Press, Castlemaine, Victoria, 2007. *(Excellent, with unusual information; by an Australian dowser.)*

—, *Stone Age Farming — Eco-Agriculture for the 21st Century,* Python Press, Castlemaine, Victoria, 2001. *(Permaculture, paramagnetism, dowsing, round towers and ancient technology.)*

Myneni, R.B., Negrón Juárez R.I., Goulden, M.L., Fu, R., Bernades, S., & Ga, H. 'An empirical approach to retrieving monthly evapotranspiration over Amazonia,' *International Journal of Remote Sensing* 2008, 29, pp. 7045–63.

Narby, Jeremy, *Intelligence in Nature,* Tarcher/Penguin, New York, 2005.

National Geographic Magazine, 'Water — Our Thirsty World,' *special issue,* April 2010.

Okri, Ben, 'The Moral Bankruptcy of our Civilization,' S&MN *Review,* no. 98, 2008.

Ostrander, Sheila & Schroeder, Lynn, *Psychic Discoveries behind the Iron Curtain,* Prentice-Hall, Englewood Cliffs, NJ, 1970.

Ott, John, *Health and Light,* Devin-Adair, Greenwich CT, 1973.

Pogacnik, Marko, *Healing the Heart of the Earth — Restoring the Subtle Levels of Life,* Findhorn Press, 1998.

—, & Werner, Karin, *Nature Spirits & Elemental Beings — Working with the Intelligence in Nature,* Findhorn Press, 1997.

Raethjen, Paul, *Dynamics of Cyclones,* Leipzig, 1953.

Ryrie, Charlie, *The Healing Energies of Water,* Gaia, London, 1998. *(Well informed and beautifully illustrated.)*

Safina, Carl, 'For Evolution to Live, Darwin Must Die,' *Observer,* 8 March, 2009.

Sahtouris, Elizabet, *Earth Dance – Living Systems in Evolution,* iUniverse.co, 2000.

—, 'Discovering Nature's Secrets of Success — A Potential Future for a Global Family,' S&MN *Review,* no. 89.

Schauberger, Viktor, *The Water Wizard (Eco-Technology,* vol. 1), Gateway, Bath, 1998.

—, *Nature as Teacher (Eco-Technology,* vol. 2), Gateway, Bath, 1998.

—, *Fertile Earth (Eco-Technology,* vol. 3), Gateway, Bath, 2000.

—, *Energy Evolution (Eco-Technology,* vol. 4), Gateway, Bath, 2000.
—, *Unsere Sinnlose Arbeit,* (*Our Senseless Toil*), Krystall 1933. Rev. ed. Schauberger-Archiv & J.Schauberger Verlag, Bad Ischl, 2003.
Schiff, Michel, *The Memory of Water,* Thorsons, London, 1995. *(Account of Jacques Benveniste's research.)*
Schmauss, August, 'Biologische Gedanken in der Meteorologie,' *Forschungen und Fortschritte,* vol. 21, 1945.
Schulz, Andreas, *Water Crystals — Making the Quality of Water Visible,* Floris Books, Edinburgh, 2005.
Schwenk, Theodore, *Sensitive Chaos,* Steiner Books, London, 1965. *(A classic on energy in water.)*
— & Schwenk, Wolfram, *Water — The Element of Life,* Anthroposophic Press, New York, 1989.
Schwenk, Wolfram (ed), *The Hidden Qualities of Water,* Floris Books, Edinburgh, 2007.
Seamon, David & Zajonc, Arthur (eds.), *Goethe's Way of Science – a Phenomenology of Nature,* SUNY Press, New York, 1998.
Sheldrake, Rupert, *The Rebirth of Nature — The Greening of Science and God,* Rider, London, 1990.
Somers, Barbara (edited by Hazel Marshall), *The Fires of Alchemy – A Transpersonal Viewpoint,* Archive, Dorset, 2004.
—, with Ian Gordon-Brown (edited by Hazel Marshall), *Journey in Depth — A Transpersonal Perspective,* Archive, Dorset, 2002.
Stevens, Peter, *Patterns in Nature,* Penguin, London, 1974.
Stewart, Iain, *How Earth Made US,* Five-part BBC science documentary, 2010.
—, & Lynch, John, *Earth, The Power of the Planet,* Six-part BBC documentary, with BBC Books, London, 2007. *(Background to the TV series — inspiring and informative; gripping photography.)*
Stone, Robert, *The Secret Life of Your Cells,* Whitford, 1982.
Tame, David, *The Secret Power of Music,* Inner Traditions, Rochester, VT.
Thomas, Lewis, *The Lives of a Cell — Notes of a Biology Watcher,* Viking, New York, 1974. *(A classic.)*
Thomas, Pat, 'Behind the Label' (bottled water), *The Ecologist,* Sep 2007.
Thomson, C. Leslie, *Water and Nature Cure,* Thomson-Kingston, Edinburgh, 1955 & 1970.
Tompkins, Peter & Bird, Christopher, *The Secret Life of Plants,* Harper & Row, New York, 1973. *(A classic.)*
—, *Secrets of the Soil,* Harper & Row, New York, 1978.
Treven, Michael & Talkenberger, Peter, *Environmental Medicine* (n.d.) *(Quoting Prof. Wolfgang Ludwig.)*
Van der Kroon, Coen, *The Golden Fountain — The Complete Guide to Urine Therapy,* Gateway, Bath, 1992.
Wales, Charles, Prince of, 'Restoring Harmony and Connection: Inner and Outer,' S&MN *Review,* no. 98.
Wilkens, Andreas, Jacobi, Michael, & Schwenk, Wolfram, *Understanding Water,* Floris Books, Edinburgh.
Wilkes, John, *Flowforms — The Rhythmic Power of Water,* Floris Books, Edinburgh, 2005.
Wilkinson, Richard & Pickett, Kate, *The Spirit Level — Why more equal Societies almost always do Better,* Penguin, London, 2009.
Wright, Machaelle, *Co-Creative Science — A revolution in science providing real solutions for everyday's health and environment,* Perelandra, Jeffersonton VA, 1997.

Links and Resources

Oceanic and Climate Change Research

Although the oceans control the world's weather and climate and contain 90% of the Earth's biomass, our knowledge about them is pitifully small. They hold the key to understanding climate change. Some good sources of information are:

Woods Hole Oceanographic Institute, Woods Hole, Massachusetts, USA. The best known ocean research centre in North America, which first drew attention about six years ago to the slowing down of the Gulf Stream

Climate Change in the Arctic: European countries have cooperated to fund important research: for instance, the Damocles Project (www.damocles-eu.org)

National Oceanographic Centre, University of Southampton, is the UK's focus on research on the oceans (www.noc.soton.ac.uk)
Hadley Centre, Exeter, is the UK's influential centre for climate change research; part of the government's Meteorological Office (www.metoffice.gov.uk/climatechange)

Water Treatment

For more information on water quality, practical suggestions, water filters & quantum water products:— see www.Sulis-health.co.uk

Water's Chemistry

There are good overviews at:

1. Jill Granger's pages at http://witcombe.sbc.edu/water/chemistrystructure.html
2. www.filtersfast.com/Water-Chemistry.asp
3. Martin Chaplin's excellent pages on water's structure and science: www.Isbu.ac.uk/water

General

The Institute of Science in Society (ISIS) was founded in 1998 by Dr Mae-Wan Ho and her husband, Peter Saunders, Professor of Applied Mathematics at King's College, University of London. A small team of dedicated pioneers supervises their pivotal research, lectures widely internationally, organizing and participating in conferences, publishing books and their bi-monthly illustrated magazine *Science in Society*. It is the only truly independent institute of its kind in the world.

If you want to be well-informed on the big issues of the day — renewable energy sources, climate change, food and energy issues, genetic engineering, the new biology and quantum physics and so on, their excellent website (www.i-sis.org.uk) is a reliable source of independent information. Their funding is by private donation. To my mind, they deserve our support more than many of the environmental charities (any money you donate them will go much further).

Viktor Schauberger website: www.Schauberger.co.uk

Author's website: www.AlickBartholomew.co.uk

Index